# Springer Theses

Recognizing Outstanding Ph.D. Research

## Aims and Scope

The series "Springer Theses" brings together a selection of the very best Ph.D. theses from around the world and across the physical sciences. Nominated and endorsed by two recognized specialists, each published volume has been selected for its scientific excellence and the high impact of its contents for the pertinent field of research. For greater accessibility to non-specialists, the published versions include an extended introduction, as well as a foreword by the student's supervisor explaining the special relevance of the work for the field. As a whole, the series will provide a valuable resource both for newcomers to the research fields described, and for other scientists seeking detailed background information on special questions. Finally, it provides an accredited documentation of the valuable contributions made by today's younger generation of scientists.

## Theses are accepted into the series by invited nomination only and must fulfill all of the following criteria

- They must be written in good English.
- The topic should fall within the confines of Chemistry, Physics, Earth Sciences, Engineering and related interdisciplinary fields such as Materials, Nanoscience, Chemical Engineering, Complex Systems and Biophysics.
- The work reported in the thesis must represent a significant scientific advance.
- If the thesis includes previously published material, permission to reproduce this must be gained from the respective copyright holder.
- They must have been examined and passed during the 12 months prior to nomination.
- Each thesis should include a foreword by the supervisor outlining the significance of its content.
- The theses should have a clearly defined structure including an introduction accessible to scientists not expert in that particular field.

More information about this series at http://www.springer.com/series/8790

Stefanie Czischek

# Neural-Network Simulation of Strongly Correlated Quantum Systems

Doctoral Thesis accepted by
Heidelberg University, Heidelberg, Germany

 Springer

*Author*
Dr. Stefanie Czischek
Kirchhoff-Institute for Physics
Heidelberg University
Heidelberg, Germany

*Supervisor*
Prof. Thomas Gasenzer
Kirchhoff-Institute for Physics
Heidelberg University
Heidelberg, Germany

ISSN 2190-5053          ISSN 2190-5061   (electronic)
Springer Theses
ISBN 978-3-030-52717-4          ISBN 978-3-030-52715-0   (eBook)
https://doi.org/10.1007/978-3-030-52715-0

This Springer imprint is published by the registered company Springer Nature Switzerland AG
The registered company address is: Gewerbestrasse 11, 6330 Cham, Switzerland

# Supervisor's Foreword

'Nature isn't classical, dammit, and if you want to make a simulation of nature, you'd better make it quantum mechanical, and by golly it's a wonderful problem, because it doesn't look so easy.'—these famous words by Richard Feynman, spoken almost 40 years ago, are often cited as a kind of midwifery for the quantum-simulation idea. During the past two decades, much experimental effort has been invested in refining the control and observation of microscopic quantum systems with many coupled degrees of freedom. The goal of this is to eventually realize universal quantum simulator platforms of the kind Feynman has mentioned in his keynote speech at the 1981 MIT conference on physics and computation (reproduced in reference [1] of the Introduction Chap. 1). In this area of quantum research, though, it is sometimes overlooked, that Feynman dedicated a large, actually, the largest part of his talk to the question of whether quantum physics can be simulated with *classical* computers, and to the difficulties and challenges connected with that. He put forward that simulating physics with computers 'scem[s] [...] to be an excellent program to follow out' which, however, needed 'full attention and acceptance' of 'the challenge of explaining quantum mechanical phenomena'.

Another thriving line of research, also over the past two decades, has given birth to so-called neuromorphic hardware. Neural-network structures, implemented as analog electronic circuits by means of semiconductor chip technology, are being developed toward capturing the functionality of the brain. With their high non-local connectivity between the neuron degrees of freedom, artificial neural nets have now come into focus for simulating static and dynamical characteristics of quantum systems. Neuromorphic realizations are particularly promising for this, as they are known to be fast in sampling and efficient in relation to their energy consumption. And they come close to what Feynman might have envisaged decades ago—when he alluded to, not strictly locally interconnected, cellular automata (see reference [1] of the Introduction Chap. 1).

So, is there a path to quantum simulation with neuromorphic hardware? When Stefanie Czischek began her doctoral research in my group in 2016, there had been barely any signs that the idea of using artificial neural networks and neuromorphic chip technology for simulating quantum physics would gain such a momentum within a short time. With her strong interest in developing computational methods for studying fundamental physics problems, she entered this new field and, adding to her experiences in quantum physics, quickly became a knowledgeable expert on machine learning techniques and bio-inspired neuromorphic networks amidst the quantum world. In her Ph.D. thesis, 'Neural-network simulation of strongly correlated quantum systems', Dr. Czischek has laid a foundation for the realization of quantum simulation on neuromorphic hardware.

It is often not straightforward to identify aspects of an observable which are unequivocally "quantum", in the sense that they require quantum physics to be there in the first place. To find out what these aspects are, it is useful to be able to benchmark a classical approximation with an exact quantum solution. For quantum spin systems, which are the first object of the neural-network implementations, Dr. Czischek demonstrates the so-called discrete Truncated Wigner approximation to exhibit one important challenge of classically simulating strong quantum correlations near a quantum critical point: It is often the extraordinary, e.g., a particularly small probability of a certain configuration, with which she distinguishes a quantum from a classically correlated system, and the simulating machine needs to make sure that the respective event just does not happen. With her in-depth study of three different approaches to representing quantum states with artificial neural networks, she accepts this and many other demanding challenges. While, in the first two approaches, she focuses on a direct representation of the complex-valued quantum amplitudes by means of Boltzmann-machine-type network architectures, in the third one she makes use of a formulation of maximally Bell-entangled quantum states in terms of positive-definite probability distributions on a doubled phase space. A path to quantum simulation with neuromorphic hardware is now on the map.

Dr. Czischek describes in detail the necessary foundations on both sides, quantum physics, and artificial neural networks, and uses them in presenting ready-to-implement techniques for neuromorphic quantum simulators. The thesis is a valuable reference for researchers from the different communities who need to understand the respective other sides' knowledge basis to be able to collaborate on the topic. It is an exciting time to work in this field, and much progress can be anticipated when one sees the great enthusiasm with which young researchers are moving forward.

Heidelberg, Germany                                           Prof. Thomas Gasenzer
May 2020

# Abstract

Approximate simulation methods of quantum many-body systems play an important role in better understanding quantum mechanical phenomena, since due to the exponentially scaling Hilbert space dimension these systems cannot be treated exactly. However, a general simulation approach is still an outstanding problem, as all efficient approximation schemes turn out to struggle in different regimes. Hence, a detailed knowledge about the limitations is an important ingredient to approximation methods and requires further studies. In this thesis, we consider two simulation approaches, namely the discrete truncated Wigner approximation as a semi-classical phase-space method, and a quantum Monte Carlo method based on a quantum state parametrization via generative artificial neural networks. We benchmark both schemes on sudden quenches in the transverse-field Ising model and point out their limitations in the quantum critical regime, where strong long-range interactions appear. Furthermore, we study the combination of the quantum state representation in terms of artificial neural networks with the neuromorphic chips present in the BrainScaleS group at Heidelberg University. The goal of this combination is to simulate entangled quantum states on a classical analog hardware. We then expect a more efficient way to approximately simulate quantum many-body systems by overcoming the limitations of conventional computation architectures, as well as further insights into quantum phenomena.

**Publications Related to This Thesis**

- S. Czischek, M. Gärttner, and T. Gasenzer, Physical Review B **98**, 024311 (2018), *"Quenches near Ising quantum criticality as a challenge for artificial neural networks"*.

- S. Czischek, M. Gärttner, M. Oberthaler, M. Kastner, and T. Gasenzer, Quantum Science and Technology **4**, 014006 (2018), *"Quenches near criticality of the quantum Ising chain–power and limitations of the discrete truncated Wigner approximation"*.

- S. Czischek, T. Gasenzer, M. Gärttner, Physical Review B **100**, 195120 (2019), *"Sampling scheme for neuromorphic simulation of entangled quantum systems"*.

# Acknowledgements

I would like to thank all the people who contributed in making this thesis possible and in making my Ph.D. time such a great experience.

First of all, I am very thankful to Prof. Thomas Gasenzer and Dr. Martin Gärttner for the great supervision and for giving me the opportunity to work on this interesting topic, which I did with great excitement. I enjoyed the time in the "Far-from-equilibrium quantum dynamics" group and am grateful for all the things I learned and experienced during the last years.

Furthermore, I thank Prof. Selim Jochim for kindly agreeing to referee this thesis.

For fruitful discussions, I am thankful to Prof. Markus Oberthaler, who always submitted great ideas by looking at things from a different point of view. For nice and valuable collaborations, I thank Prof. Michael Kastner and Prof. Jan M. Pawlowski.

My special thanks go to Andreas Baumbach for being so patient in translating my ideas into the language of the neuromorphic hardware. Furthermore, I thank Lukas Kades for great discussions and brainstorming, which lead to many ideas on our joint project. For proofreading, I thank Felix Behrens, Dr. Andrea Demetrio, Aleksandr Mikheev, and Linda Shen, as well as Robert Klassert, Laurin Fischer, Timo Gierlich, and Julius Vernie for helpful feedback on the introduction. I am very grateful to the whole SynQS team for interesting discussions and for making my time at KIP so unique.

This work was supported by the state of Baden-Württemberg through bwHPC. I thank Maurits W. Haverkort for giving me access to the MLS&WISO cluster.

Very special thanks go to my parents and my brother for supporting me in everything I do. I am deeply thankful to my husband Jan for always being by my side.

# Contents

# Chapter 1
# Introduction

In the age of quantum computers and quantum simulators appearing in the daily news, a proper understanding and controllability of quantum many-body systems forming the underlying principles becomes more and more important, albeit it is far from complete.

In 1936, Alan Turing introduced the Turing machine as a universal simulator of other existing computational models, providing insights into understanding their behavior and their underlying methods. In an analogous way, people nowadays investigate the use of ordinary computers for simulating quantum mechanical systems, expecting further insights into the so-far not well understood quantum world. However, Richard Feynman stated already at the 1981 MIT conference on physics and computation that there exist fundamental difficulties in setting up simulations of quantum systems on classical computers. Even in these early days he suggested to build computers based on quantum mechanics, as reproduced in [1]. This statement got further approved since then, as even until today quantum many-body systems cannot be simulated efficiently on a classical computer. Here the term efficiently refers to the computational costs of the simulation method scaling polynomially in the problem size, rather than super-polynomially or even exponentially [2].

A first ansatz to overcome this problem are experimental quantum simulators, where a well-understood and controllable experiment is used to simulate the behavior of a more complex quantum system [3]. This enables observations of such complex quantum systems and even of dynamics happening therein, for example after driving the system out of equilibrium. Different ways and ideas to build such quantum simulators exist and have been implemented. Considering spin models, they can, for example, be simulated by controlled experiments based on trapped ions [4–7], Rydberg atoms [8–11], polar molecules [12], or ultracold atoms in continuum and lattice traps [13–18].

To ensure that such quantum simulators are describing the desired model, they can be benchmarked on approximate numerical simulations in regimes where those are still known to work well, as it is done in [15] for strongly interacting Bose gases. This

S. Czischek, *Neural-Network Simulation of Strongly Correlated Quantum Systems*,
Springer Theses, https://doi.org/10.1007/978-3-030-52715-0_1

emphasizes the importance of numerical methods to efficiently approximate quantum systems, at least in some regimes. To develop and evolve efficient numerical methods which approximately simulate the behavior of a quantum system with high accuracy, one can start by considering simple systems, specifically spin-1/2 systems. Those consist of binary "particles" which can take two possible states, often referred to as spin up and spin down. Even though having such a simple structure they still show fundamental physical phenomena. They appear in many condensed-matter problems since they can be related to magnetism and they can be considered as a limiting case of bosons on a lattice (Bose–Hubbard models) for certain parameter regimes.

The crucial point making these systems interesting is the analytical solvability which is still given for some of them, especially for small system sizes, and hence enables a suitable benchmark for approximate numerical simulation methods. Furthermore, spin-like models appear in many different fields and hence spins have various interpretations. Physical spins are defined as an intrinsic angular momentum of particles, where for example an electron is a spin-1/2 particle. Often also polarizations of photons, or particles in a ground or excited state, such as Rydberg atoms, are referred to as pseudo-spins. These describe two-level systems and hence are binary. Coming back to quantum computers, their binary units are known as qubits (quantum-bits), so that in principle quantum computers are systems of many spin-1/2 particles. This emphasizes the urge for efficient numerical simulation methods of quantum spin-1/2 models to gain further insights into quantum computation.

However, even though being urgent, efficient simulations of quantum spin systems and especially of their dynamics are still an outstanding problem due to the Hilbert space dimension scaling exponentially with system size, as illustrated in Fig. 1.1. This phenomenon is known as the curse of dimensionality. Due to superpositions being present in quantum mechanics, an exponentially growing Hilbert space dimension leads to exponentially growing computational costs to exactly simulate the system. Thus, multiple ways to approximate spin-1/2 models have been introduced. For some models efficient exact (within numerical errors) simulation methods exist, as used in, e.g., [19], but there is no general ansatz.

Many of these approximation methods belong to the class of tensor-network states, probably the most famous one being the (time-dependent) density-matrix renormalization group ([t]DMRG), which is based on expressing the quantum states in terms of matrix products (matrix product states [MPS]) [20]. It has been introduced in [21], with an extension to simulate dynamics in [22], and a detailed overview is given in [23, 24]. This method is applied frequently, especially to simulate dynamics in quantum systems [25–27]. Several extensions have been proposed, for example [28, 29].

The basic ansatz of the DMRG method is the observation that for many quantum states not the whole Hilbert space needs to be considered. Instead, the Hilbert space contains a huge number of configurations of which only a small subgroup yields states the quantum system can be found in with significant probabilities. To those we refer as physical states and they are illustrated on the right in Fig. 1.1. The remaining states (dimensions) can be neglected when approximating the system. In some cases

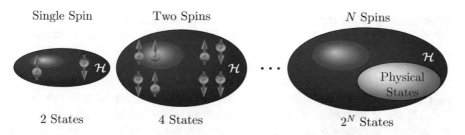

**Fig. 1.1** Schematic illustration of the curse of dimensionality. The Hilbert space of a quantum spin-1/2 system contains all states the system can be found in and is illustrated by the blue regimes (indicated by $\mathcal{H}$). The individual spin particles are depicted by the arrows, where the directions (up and down) illustrate the two possible states the particle can take. For a single spin, the Hilbert space contains two possible basis states (depicted on the left), while it already contains four possible basis states for a system of two spin particles (depicted in the center). This shows the exponential scaling of the Hilbert space dimension, so that for the general case of $N$ spins $2^N$ possible basis states exist (depicted on the right, states are not shown anymore). The subset of physical states is illustrated for this general case. Those are the states the system is found in with significant probabilities. The dimension of this subspace is reduced compared to the full Hilbert space dimension, since several states, such as, e.g., strongly entangled states, are not contained. If the dimension of this subspace scales polynomially with the system size, which is true in some cases, the full system can be approximated efficiently by only considering those physical states and neglecting the remaining ones

the number of physical states grows polynomially with the system size, enabling an efficient way to approximately simulate the system with good accuracy [23, 24].

The problem is hence shifted from solving the model with exponentially growing dimensionality to finding the relevant states. The DMRG approaches this problem using a variational ansatz and truncating the density matrix at a fixed order. By varying the order of truncation, the so-called bond dimension, it can be checked if the approximation still represents the model accurately. If this is true, increasing the bond dimension further does not yield any changes. Thus, the (t)DMRG is a controllable approximation method for ground states and dynamics in spin-1/2 systems [23, 24]. However, such MPS-based methods are limited to systems in one spatial dimension and they only work efficiently if the system is not strongly entangled. Otherwise the bond dimension to reach convergence still scales exponentially with the system size. A discussion on the applicability is given in [23, 24].

Another class of approximate simulation methods are phase-space methods, like the truncated Wigner approximation (TWA), which is a semi-classical simulator, as discussed in detail in [30, 31]. These methods are used to describe quantum dynamics, for example after quenching a system out of equilibrium. They are based on applying classical equations of motion to initial quantum states sampled in phase space. By averaging the outcomes of multiple classical trajectories resulting from different sampled initial states, the dynamics of the quantum system are approximated in a semi-classical fashion.

The applicability of those phase-space methods is not limited in the spatial system dimension and they have been found to perform well at short time scales, while struggling at strong interacting regimes and long times. Being semi-classical, they might not capture all quantum mechanical effects. However, they are applied commonly, for example to simulate bosonic systems out of equilibrium [14, 32, 33].

The TWA has been extended to simulate discrete quantum systems, such as spin-1/2 systems, using a discrete phase space as proposed in [34] with an extension in [35]. This discrete truncated Wigner approximation (dTWA) has been introduced in [36, 37] and a truncation at higher orders has been studied in [38]. Furthermore, the method has been widely extended, for example to derive a general discretization of quasi-probability distributions [39], or to simulate scattering of light in atomic clouds [40]. Moreover, a generalized dTWA has been introduced recently in [41], enabling semi-classical simulations not only of spin-1/2 particles, but arbitrary spin systems. A cluster TWA has been introduced in [42], which uses a generalization of the TWA to a higher-dimensional phase space.

In contrast to the TWA, the dTWA has been found to perform well even at longer time scales and in strongly correlated systems [38]. Showing such a remarkable accuracy, the dTWA is used frequently, for example to simulate Heisenberg spin systems [43], spin-boson models [44], Rydberg systems [45], or lattices of dipolar molecules [46]. It has furthermore been used to investigate many-body localization and thermalization [47]. However, it is not yet clearly understood how accurately the dTWA can capture quantum effects such as quantum entanglement, or how well it performs in quantum critical regimes close to a quantum phase transition. Therefore, in this thesis we benchmark the dTWA on sudden quenches in the transverse-field Ising model (TFIM), which defines a quantum phase transition. We point out the limitations of the simulation method for quenches close to the quantum-critical point in Chap. 4 according to [48].

A third famous family of approximate simulation schemes for quantum many-body systems are quantum Monte Carlo (QMC) methods, which are commonly used to study fermion systems where semi-classical approximation methods are in general not applicable. Detailed reviews for QMC methods are given in [49, 50]. Those methods are based on considering the basis states of the quantum system and the probability distribution underlying those, which is defined via the squared wave function evaluated at the individual basis states. This probability distribution provides a measure for the importance of the basis states to describe the full system. Hence, a Monte Carlo sampling scheme (importance sampling) can be used to draw the physical states according to the underlying distribution while neglecting the remaining ones. Expectation values of observables can be approximated by evaluating them at the sampled states and normalizing the outcome, where the complex phases of the wave function need to be considered.

As it is still exponentially hard to calculate the wave function, this is as well considered in an approximate manner. To do so, a suitable parametrization of the coefficients in a basis expansion of the state vector can be chosen. With the help of a variational ansatz this parametrization can be adapted such that it represents the desired spin state [51]. Multiple possible choices for parametrizations exist, where it depends on the considered model which of them is most suitable. Moreover, those variational QMC schemes can be extended to simulate dynamics in quantum

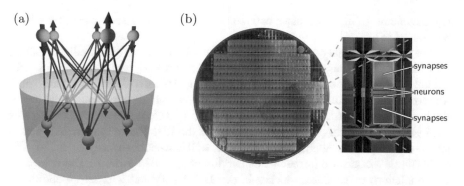

**Fig. 1.2 a** Illustration of a restricted Boltzmann machine (RBM). The network contains a layer of binary visible neurons (orange) and binary hidden neurons (green, in shaded region) with arrows denoting the two possible states (up and down). All-to-all connections are applied between the two layers. The strengths of the connections are variational parameters and define a Boltzmann distribution underlying all neurons [54]. By adjusting the connection strengths, any joint probability distribution underlying the visible neurons can be modeled given a sufficient amount of hidden neurons [55]. By identifying the visible neurons with particles in a spin-1/2 system, the probability distribution underlying spin configurations in a specific state can be modeled, providing a way to sample physical states from the Hilbert space [56, 57]. We analyze this ansatz in more detail in this thesis. Figure adapted from [58]. **b** Photograph of a wafer containing neuromorphic chips in the BrainScaleS system at Heidelberg University. This mixed-signal neuromorphic hardware can be used to emulate an RBM via physically implemented spiking neurons and can efficiently draw samples from the Boltzmann distribution underlying the visible and hidden neurons. The wafer consists of 48 units (reticles) of 8 HICANN (High Input Count Analog Neural Network) chips each. Every chip contains 512 neurons and one such chip is enlarged on the right. The neurons on the chips are connected via synapses found in the two large regions and the individual units are connected externally [59]. In this thesis we analyze the way towards simulating entangled quantum systems with the help of such neuromorphic hardware, from which we expect an immense speedup in computation time. The images were provided by A. Baumbach (BrainScaleS, Heidelberg University) and the figure setup is adapted from [59]

systems, yielding the time-dependent variational Monte Carlo ansatz, as introduced and applied in [52, 53].

During the last years, a parametrization of the state vector based on artificial neural networks, specifically restricted Boltzmann machines (RBM), has been introduced and caught great attention [56, 57]. An RBM is a two-layer neural network with one visible and one hidden layer and only inter-layer interactions, as illustrated in Fig. 1.2a [54]. The visible and hidden neurons follow a Boltzmann distribution defined by the strengths of the connections in the network, which can be trained to represent the distribution underlying the basis states in a quantum spin-1/2 system. This parametrization has been studied and applied frequently for pure [60–63] and mixed states [64–68], where the representation of quantum states is found from given data [64, 69–71] or via a variational ansatz [57, 65–68, 72] for either purely positive states or with different ways to include complex phases. Furthermore, different representations of quantum states based on feed-forward neural networks have been introduced in [73, 74].

The representational power of the RBM parametrization has been studied in detail, for example how accurately quantum entangled states [62], topological states [75], strongly correlated systems [63], or chiral topological phases [76] can be represented. Moreover, the effect of going to deep neural networks on the representational power has been studied in [61, 77, 78] and an extension to different network architectures has been introduced in [79]. The connection of the RBM-based parametrization to tensor-network states caught great attention as well and has been analyzed in [80–83]. In this thesis we perform further benchmarking of the RBM parametrization method for simulations of sudden quenches in the TFIM with and without an additional longitudinal field. We point out the limitations of the ansatz for quenches into regimes of large correlation lengths in Chap. 5 according to [58].

While the applicability of such QMC-based methods is not limited in the system dimension, problems appear in capturing the complex phases of the basis-expansion coefficients. As these cannot be directly considered in the sampling scheme, they need to be treated separately. Besides various existing specific methods to deal with the phases for individual parametrizations, a common general ansatz is to sample from the distribution given by the real amplitudes of the basis-expansion coefficients. The observables evaluated at the sampled states are then explicitly multiplied with the corresponding phases, which can be extracted from the parametrization.

This so-called (phase) reweighting approach can be computationally expensive depending on the chosen parametrization ansatz. Additionally, it leads to the famous sign problem for states with strongly fluctuating phase factors. Due to these fluctuations, an exponential scaling of the sample size necessary to reach a certain accuracy for increasing system sizes arises, rendering the ansatz inefficient as discussed in [84–87]. Several approaches have been introduced to deal with the sign problem [88–90]. In the case of the RBM parametrization the phases can be included by choosing complex connection strengths in the neural network. This circumvents the sign problem but causes limitations in changing the network structure and hence in increasing the representational power of the parametrization ansatz [57]. In this thesis we benchmark the sampling via the (phase) reweighting scheme from a complex-valued RBM parametrizing entangled quantum states. We show in Chap. 6 according to [91] that the necessary number of samples to reach a desired accuracy indeed scales exponentially with the system size for certain states, indicating the existence of the sign problem.

A further limitation of (variational) QMC-based methods is that it is not clear if any arbitrary wave function can be represented by the chosen parametrization. In the case of the RBM-based ansatz, its representational power can be increased by adding more hidden neurons, but this also increases the computational costs. While it has been shown that a real-valued RBM can approximate any joint probability distribution underlying the visible neurons arbitrarily well given enough hidden neurons [55], it is not clear if this is also true for an RBM with complex connections. Furthermore, if the necessary number of hidden neurons scales exponentially with the system size, the QMC method is not efficient anymore.

(a)

(b)

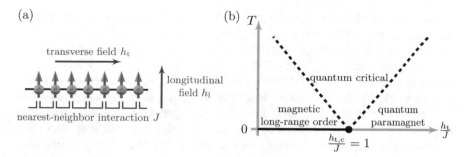

**Fig. 1.3** **a** Schematic illustration of the one-dimensional Ising model with nearest-neighbor Ising interactions of strength $J$ in a transverse field $h_t$ and a longitudinal field $h_l$. The arrows denote the quantum spin-1/2 particles in the basis where interactions are applied and their orientation denotes the two possible states (up and down). The system is shown in a fully polarized state as an example. **b** Phase diagram of the transverse-field Ising model with $h_l = 0$ [92]. The model undergoes a quantum phase transition at temperature $T = 0$ and $h_t/J = 1$. As the model is integrable, we use it in this thesis to benchmark approximate simulation methods of quantum spin-1/2 systems in the vicinity of the quantum critical point, i.e., we only consider the case of zero temperature

Comparing the three introduced simulation approaches, MPS based methods are best understood and controllable. For those it can be checked via increasing the bond dimension whether the simulations have converged or not. If they have converged, they describe the exact solution accurately. Such a controlled approximation error is not given for phase-space and QMC methods, so that these still require further benchmarking.

A standard model to consider when benchmarking approximate simulation methods for spin-1/2 systems is the one-dimensional TFIM. This is a spin chain with nearest-neighbor Ising interactions in a transverse field, as illustrated in Fig. 1.3a. The model describes a quantum phase transition between a para- and a ferromagnetic phase at a quantum critical point, see Fig. 1.3b, and is integrable, meaning that it can be solved analytically. Thus, it is an exactly solvable system on which simulations can be benchmarked at quantum criticality. Properties of the model as well as its phase diagram have been studied in detail on the analytical solution [92–94]. Dynamics after sudden changes (quenches) in the transverse field have been analyzed a lot in, e.g., [95–100]. Thus, the TFIM is a commonly chosen model to benchmark simulation approaches such as phase-space or QMC methods for arbitrary system sizes also in (quantum) critical regimes.

While huge effort is expended in finding and benchmarking such approximate simulation methods of quantum systems, a further approach to overcome the limitations of classical computers in quantum simulations is to use different computing devices, exceeding the von-Neumann architecture. Quantum computers are obviously not an option here as the problems arise due to the motivation of getting further insights into quantum computation. Instead, more architectures exist, such as neuromorphic chips. Those emulate biological neural networks on physical models of neurons connected via synapses, implemented by mixed-signal microelectronics, as depicted in Fig. 1.2b [101–104].

Given the RBM-based parametrization, it seems to be a reasonable motivation to use such neuromorphic chips for simulations of quantum many-body systems. Furthermore, the BrainScaleS group at Heidelberg University is evolving and using such a neuromorphic hardware, from which the photographs in Fig. 1.2b are taken, with a general overview given in [102]. This enables us to directly implement a suitable algorithm. The hardware can sample efficiently from a Boltzmann distribution, as has been shown in [59, 103, 105]. For this task a speedup of at least one order of magnitude is expected compared to sampling on a classical computer, with about three orders of magnitude less energy consumption [106]. Thus, from a combination of quantum systems and neuromorphic chips we expect an efficient way to sample quantum states in a QMC-based approach if the underlying distribution can be expressed in a Boltzmann form. Furthermore, by emulating strongly entangled states on the analog hardware we expect further insights into quantum mechanical effects.

However, as wave functions of quantum systems are complex-valued quantities, a translation to the real-valued signals flowing through a neuromorphic hardware causes troubles, since it is not clear how to include the complex phases. Thus, a direct implementation of the RBM-based parametrization with complex-valued connections on the neuromorphic hardware is not possible, even though it sounds like a trivial task. This problem can be overcome by using the general reweighting ansatz, where the real amplitudes of the basis-expansion coefficients are parametrized by the RBM structure containing the real parts of the connection strengths. The imaginary parts provide the complex phases which are used to reweight the observables when calculating expectation values. We discuss and analyze this ansatz further in Chap. 6 based on [91]. Alternatively, the parametrization can be modified in a way to match the network structure present on the neuromorphic chips, meaning that the quantum system is represented by a purely real-valued RBM. This modification we discuss further in Chap. 7. With this, we provide and analyze two ways towards simulating quantum spin-1/2 systems with the help of physically implemented neurons in the neuromorphic hardware in this thesis.

The thesis is structured as follows. In Chap. 2 we give a more detailed background on quantum mechanics and spin-1/2 systems, as well as on the transverse-field Ising model and quantum entanglement. For this we take the strongly entangled Bell state and its generalization to larger spin systems, the Greenberger–Horne–Zeilinger (GHZ) state, as examples. Those states have been introduced and discussed in [107–110], respectively. We discuss the analytical solution of the TFIM according to [92–100] and consider the (t)DMRG method according to [23, 24]. Then exact diagonalization, a simulation method for small system sizes, is considered in detail. With this we gather all ingredients to enable a systematic benchmarking of approximate simulation methods.

In Chap. 3 we discuss the background of general artificial neural network setups and their applications in machine learning. For the profit of better understanding the RBM-based parametrization of quantum states, we provide a general overview containing approaches which are used in later chapters of this thesis. This chapter furthermore builds the foundations of adapting the parametrization ansatz to enable an implementation on the neuromorphic hardware, which is also discussed thoroughly within this chapter. Additionally, an overview of machine learning applications in physics is given.

Chapter 4 introduces the dTWA in detail according to [36–38] and shows benchmarks on sudden quenches in the TFIM. The quenches start from a fully polarized initial state and go to different transverse fields in the para- and ferromagnetic phase, as well as into the vicinity of the quantum critical point. For quenches close to the quantum phase transition we find strong long-range correlations and volume-law entanglement. We show that these are limiting factors for the dTWA at late times, while the short-time behavior is still captured well.

In Chap. 5 we analogously benchmark the RBM-based QMC method on sudden quenches in the TFIM after having introduced the setup and properties of the method in detail according to [56, 57]. We again consider sudden quenches from a fully polarized state into both phases and close to the quantum phase transition and provide direct comparisons to dTWA results. Furthermore, we study the representational power of the parametrization ansatz by varying the number of hidden neurons in the RBM. We again find limitations close to the quantum critical point in the regime of strong correlations and volume-law entanglement. However, we find convergence when increasing the number of hidden neurons up to the size of the Hilbert space dimension, providing again an exponential scaling with system size. We also add a longitudinal field to the TFIM, as illustrated in Fig. 1.3a, making the model non-integrable. We then benchmark the dTWA and the RBM-based QMC ansatz on exact diagonalization and tDMRG results. This yields analogous observations as in the TFIM.

Thus, we show that the dTWA, as well as the RBM-based QMC ansatz do not exceed the limitations of MPS-based methods which struggle in regimes of strong entanglement due to an exponentially growing bond dimension. However, in contrast to MPS-based methods, these two approaches are applicable to systems in higher spatial dimensions. There they are expected to perform analogously well in regimes with weak (quantum) correlations.

In Chap. 6 we approach the task of using neuromorphic chips to make simulations of spin systems more efficient. We apply the reweighting scheme to deal with the complex phases in the RBM-based parametrization. We furthermore introduce an extension to deep neural networks to enable measurements of general operators in different cartesian bases by observing the visible layer of the network. We show that with this ansatz the ground state of the TFIM at the quantum critical point can be represented with good accuracy and expectation values of operators in the $z$-basis can be evaluated efficiently. Nevertheless, exponentially many samples are necessary to approximate measurements in the $x$-basis accurately. Moreover, when we turn to representing strongly entangled states, such as Bell and GHZ states, we experience the sign problem for measurements in both the $x$- and $z$-basis. This appears to render the simulation method inefficient and to limit its applicability to small system sizes. However, the introduced ansatz provides an efficient way to represent quantum states of small spin systems on neuromorphic chips. From this we expect that the limitations are shifted to larger system sizes due to the speedup in the sampling procedure.

Chapter 7 discusses a different ansatz of adapting the RBM-based parametrization to enable an implementation on the neuromorphic hardware. This method is based on positive-operator valued measures (POVM) and has been introduced in [64]. With the POVM ansatz, a general quantum state can be transformed into a basis where it is represented with a purely real wave function, so that all complex phases vanish. This enables a parametrization with a standard RBM as applied in machine learning tasks, which can be emulated by the neuromorphic hardware. The authors in [64] have shown already that strongly entangled states, specifically Bell and GHZ states, can be represented accurately this way. We discuss the method in detail according to [64] and train an RBM to represent a Bell state on a classical computer. Furthermore, we discuss the steps towards an implementation on neuromorphic chips in the BrainScaleS group as an outlook. With this we provide an ansatz to simulate strongly entangled quantum states with the help of classical analog hardware.

We close the thesis with a summary and conclusions in Chap. 8.

# References

1. Feynman RP (1982) Simulating physics with computers. Int J Theor Phys 21(6):467–488. https://doi.org/10.1007/BF02650179
2. Nielsen MA, Chuang IL (2010) Quantum Computation and Quantum Information: 10th, Anniversary edn. Cambridge University Press, Cambridge. https://doi.org/10.1017/CBO9780511976667
3. Ignacio Cirac J, Zoller P (2012) Goals and opportunities in quantum simulation. Nat Phys 8:264. https://doi.org/10.1038/nphys2275
4. Jurcevic P, Lanyon BP, Hauke P, Hempel C, Zoller P, Blatt R, Roos CF (2014) Quasiparticle engineering and entanglement propagation in a quantum many-body system. Nature 511:202. https://doi.org/10.1038/nature13461
5. Britton JW, Sawyer BC, Keith AC, Joseph Wang C-C, Freericks JK, Uys H, Biercuk MJ, Bollinger JJ (2012) Engineered two-dimensional using interactions in a trapped-ion quantum simulator with hundreds of spins. Nature 484:489. https://doi.org/10.1038/nature10981
6. Zhang J, Pagano G, Hess PW, Kyprianidis A, Becker P, Kaplan H, Gorshkov AV, Gong Z-X, Monroe C (2017) Observation of a many-body dynamical phase transition with a 53-qubit quantum simulator. Nature 551:601. https://doi.org/10.1038/nature24654
7. Blatt R, Roos CF (2012) Quantum simulations with trapped ions. Nat. Phys. 8:277. https://doi.org/10.1038/nphys2252
8. Schauß P, Cheneau M, Endres M, Fukuhara T, Hild S, Omran A, Pohl T, Gross C, Kuhr S, Bloch I (2012) Observation of spatially ordered structures in a two-dimensional Rydberg gas. Nature 491:87. https://doi.org/10.1038/nature11596
9. Bernien H, Schwartz S, Keesling A, Levine H, Omran A, Pichler H, Choi S, Zibrov AS, Endres M, Greiner M, Vuletić V, Lukin MD (2017) Probing many-body dynamics on a 51-atom quantum simulator. Nature 551:579. https://doi.org/10.1038/nature24622
10. Barredo D, Lienhard V, de Léséleuc S, Lahaye T, Browaeys A (2018) Synthetic three-dimensional atomic structures assembled atom by atom. Nature 561:79–82. https://doi.org/10.1038/s41586-018-0450-2
11. Günter G, Schempp H, Robert-de Saint-Vincent M, Gavryusev V, Helmrich S, Hofmann CS, Whitlock S, Weidemüller M (2013) Observing the dynamics of dipole-mediated energy transport by interaction-enhanced imaging. Science 342(6161):954–956. https://science.sciencemag.org/content/342/6161/954
12. Hazzard KRA, Gadway B, Foss-Feig M, Yan B, Moses SA, Covey JP, Yao NY, Lukin MD, Ye J, Jin DS, Rey AM (2014) Many-body dynamics of dipolar molecules in an optical lattice. Phys Rev Lett 113:195302. https://link.aps.org/doi/10.1103/PhysRevLett.113.195302

13. Braun S, Friesdorf M, Hodgman SS, Schreiber M, Ronzheimer JP, Riera A, del Rey M, Bloch I, Eisert J, Schneider U (2015) Emergence of coherence and the dynamics of quantum phase transitions. PNAS 112(12):3641–3646. http://www.pnas.org/content/112/12/3641.abstract
14. Nicklas E, Karl M, Höfer M, Johnson A, Muessel W, Strobel H, Tomkovič J, Gasenzer T, Oberthaler MK (2015) Observation of scaling in the dynamics of a strongly quenched quantum gas. Phys Rev Lett 115:245301. https://link.aps.org/doi/10.1103/PhysRevLett.115.245301
15. Trotzky S, Pollet L, Gerbier F, Schnorrberger U, Bloch I, Prokof'ev NV, Svistunov B, Troyer M (2010) Suppression of the critical temperature for superfluidity near the Mott transition. Nat Phys 6:998. https://doi.org/10.1038/nphys1799
16. Mazurenko A, Chiu CS, Ji G, Parsons MF, Kanász-Nagy M, Schmidt R, Grusdt F, Demler E, Greif D, Greiner M (2017) A cold-atom Fermi-Hubbard antiferromagnet. Nature 545:462. https://doi.org/10.1038/nature22362
17. Schreiber M, Hodgman SS, Bordia P, Lüschen HP, Fischer MH, Vosk R, Altman E, Schneider U, Bloch I (2015) Observation of many-body localization of interacting fermions in a quasirandom optical lattice. Science 349(6250):842–845. https://science.sciencemag.org/content/349/6250/842
18. Bloch I, Dalibard J, Sylvain N (2012) Quantum simulations with ultracold quantum gases. Nat Phys 8:267. https://doi.org/10.1038/nphys2259
19. Rigol M, Dunjko V, Olshanii M (2008) Thermalization and its mechanism for generic isolated quantum systems. Nature 452:854. https://doi.org/10.1038/nature06838
20. Orús R (2014) A practical introduction to tensor networks: matrix product states and projected entangled pair states. Ann Phys 349:117–158. https://www.sciencedirect.com/science/article/pii/S0003491614001596
21. White SR (1992) Density matrix formulation for quantum renormalization groups. Phys Rev Lett 69:2863–2866. https://link.aps.org/doi/10.1103/PhysRevLett.69.2863
22. Vidal G (2004) Efficient simulation of one-dimensional quantum many-body systems. Phys Rev Lett 93:040502. https://link.aps.org/doi/10.1103/PhysRevLett.93.040502
23. Schollwöck U (2011) The density-matrix renormalization group in the age of matrix product states. Ann Phys 326(1):96–192. http://www.sciencedirect.com/science/article/pii/S0003491610001752
24. Bridgeman JC, Chubb CT (2017) Hand-waving and interpretive dance: an introductory course on tensor networks. J Phys A: Math Theor, 50(22):223001. https://doi.org/10.1088%2F1751-8121%2Faa6dc3
25. White SR, Feiguin AE (2004) Real-time evolution using the density matrix renormalization group. Phys Rev Lett 93:076401. https://link.aps.org/doi/10.1103/PhysRevLett.93.076401
26. Kollath C, Läuchli AM, Altman E (2007) Quench dynamics and nonequilibrium phase diagram of the Bose-Hubbard model. Phys Rev Lett 98:180601. https://link.aps.org/doi/10.1103/PhysRevLett.98.180601
27. Sharma S, Suzuki S, Dutta A (2015) Quenches and dynamical phase transitions in a nonintegrable quantum Ising model. Phys Rev B 92:104306. https://link.aps.org/doi/10.1103/PhysRevB.92.104306
28. Daley AJ, Kollath C, Schollwöck U, Vidal G (2004) Time-dependent density-matrix renormalization-group using adaptive effective Hilbert spaces. J Stat Mech: Theory Exp 2004(04):P04005. http://stacks.iop.org/1742-5468/2004/i=04/a=P04005
29. Haegeman J, Lubich C, Oseledets I, Vandereycken B, Verstraete F (2016) Unifying time evolution and optimization with matrix product states. Phys Rev B 94:165116. https://link.aps.org/doi/10.1103/PhysRevB.94.165116
30. Polkovnikov A (2010) Phase space representation of quantum dynamics. Ann Phys 325(8):1790. https://doi.org/10.1016/j.aop.2010.02.006
31. Blakie PB, Bradley AS, Davis MJ, Ballagh RJ, Gardiner CW (2008) Dynamics and statistical mechanics of ultra-cold bose gases using c-field techniques. Adv Phys 57(5):363–455. https://doi.org/10.1080/00018730802564254
32. Karl M, Gasenzer T (2017) Strongly anomalous non-thermal fixed point in a quenched two-dimensional Bose gas. New J Phys 19(9):093014. https://doi.org/10.1088%2F1367-2630%2Faa7eeb

33. Karl M, Nowak B, Gasenzer T (2013) Universal scaling at nonthermal fixed points of a two-component Bose gas. Phys Rev A 88:063615 Dec. https://link.aps.org/doi/10.1103/PhysRevA.88.063615
34. Wootters WK (1987) A Wigner-function formulation of finite-state quantum mechanics. Ann Phys 176(1):1–21. http://www.sciencedirect.com/science/article/pii/000349168790176X
35. Wootters WK (2003) Picturing qubits in phase space. arXiv:quant-ph/0306135
36. Schachenmayer J, Pikovski A, Rey AM (2015) Dynamics of correlations in two-dimensional quantum spin models with long-range interactions: A phase-space Monte-Carlo study. New J Phys 17(6):065009. https://doi.org/10.1088/1367-2630/17/6/065009
37. Schachenmayer J, Pikovski A, Rey AM (2015) Many-body quantum spin dynamics with Monte Carlo trajectories on a discrete phase space. Phys Rev X 5:011022. https://link.aps.org/doi/10.1103/PhysRevX.5.011022
38. Pucci L, Roy A, Kastner M (2016) Simulation of quantum spin dynamics by phase space sampling of Bogoliubov-Born-Green-Kirkwood-Yvon trajectories. Phys Rev B 93(17):174302. https://link.aps.org/doi/10.1103/PhysRevB.93.174302
39. Žunkovič B (2015) Continuous phase-space methods on discrete phase spaces. EPL 112:10003. https://doi.org/10.1209/0295-5075/112/10003
40. Pucci L, Roy A, do Espirito Santo TS, Kaiser R, Kastner M, Bachelard R (2017) Quantum effects in the cooperative scattering of light by atomic clouds. Phys Rev A 95:053625. https://link.aps.org/doi/10.1103/PhysRevA.95.053625
41. Zhu B, Rey AM, Schachenmayer J (2019) A generalized phase space approach for solving quantum spin dynamics. New J Phys 21(8):082001. https://doi.org/10.1088%2F1367-2630%2Fab354d
42. Wurtz J, Polkovnikov A, Sels D (2018) Cluster truncated Wigner approximation in strongly interacting systems. Ann Phys 395:341–365. http://www.sciencedirect.com/science/article/pii/S0003491618301647
43. Babadi M, Demler E, Knap M (2015) Far-from-equilibrium field theory of many-body quantum spin systems: Prethermalization and relaxation of spin spiral states in three dimensions. Phys Rev X 5:041005. https://link.aps.org/doi/10.1103/PhysRevX.5.041005
44. Piñeiro Orioli A, Safavi-Naini A, Wall ML, Rey AM (2017) Nonequilibrium dynamics of spin-boson models from phase-space methods. Phys Rev A 96:033607. https://link.aps.org/doi/10.1103/PhysRevA.96.033607
45. Piñeiro Orioli A, Signoles A, Wildhagen H, Günter G, Berges J, Whitlock S, Weidemüller M (2018) Relaxation of an isolated dipolar-interacting Rydberg quantum spin system. Phys Rev Lett 120:063601. https://link.aps.org/doi/10.1103/PhysRevLett.120.063601
46. Covey JP, De Marco L, Acevedo ÓL, Rey AM, Ye J (2018) An approach to spin-resolved molecular gas microscopy. New J Phys 20(4):043031. https://doi.org/10.1088%2F1367-2630%2Faaba65
47. Acevedo OL, Safavi-Naini A, Schachenmayer J, Wall ML, Nandkishore R, Rey AM (2017) Exploring many-body localization and thermalization using semiclassical methods. Phys Rev A 96:033604. https://link.aps.org/doi/10.1103/PhysRevA.96.033604
48. Czischek S, Gärttner M, Oberthaler M, Kastner M, Gasenzer T (2018) Quenches near criticality of the quantum using chain-power and limitations of the discrete truncated Wigner approximation. Quantum Sci Technol 4(1):014006. http://stacks.iop.org/2058-9565/4/i=1/a=014006
49. von der Linden W (1992) A quantum Monte Carlo approach to many-body physics. Phys Rep 220(2):53–162. http://www.sciencedirect.com/science/article/pii/037015739290029Y
50. Masuo S (ed) (1986) Quantum monte carlo methods in equilibrium and nonequilibrium systems, vol 74. Solid-State sciences. Springer, Berlin, Heidelberg. https://doi.org/10.1007/978-3-642-83154-6
51. Rubenstein B (2017) Introduction to the variational monte carlo method in quantum chemistry and physics, pp. 285–313. Springer, Singapore. https://doi.org/10.1007/978-981-10-2502-0_10
52. Carleo G, Becca F, Schió M, Fabrizio M (2012) Localization and glassy dynamics of many-body quantum systems. Sci Rep 2:243. http://dx.doi.org/10.1038/srep00243

53. Carleo G, Becca F, Sanchez-Palencia L, Sorella S, Fabrizio M (2014) Light-cone effect and supersonic correlations in one- and two-dimensional bosonic superfluids. Phys Rev A 89(3):031602. https://link.aps.org/doi/10.1103/PhysRevA.89.031602
54. Hinton GE (2012) A practical guide to training restricted boltzmann machines, pp 599–619. Springer, Berlin, Heidelberg. https://doi.org/10.1007/978-3-642-35289-8_32
55. Le Roux N, Bengio Y (2008) Representational power of restricted Boltzmann machines and deep belief networks. Neural Comput 20(6):1631–1649. https://doi.org/10.1162/neco.2008. 04-07-510
56. Torlai G, Melko RG (2016) Learning thermodynamics with Boltzmann machines. Phys Rev B 94:165134. https://link.aps.org/doi/10.1103/PhysRevB.94.165134
57. Carleo G, Troyer M (2017) Solving the quantum many-body problem with artificial neural networks. Science 355(6325):602–606. http://science.sciencemag.org/content/355/6325/602
58. Czischek S, Gärttner M, Gasenzer T (2018) Quenches near using quantum criticality as a challenge for artificial neural networks. Phys Rev B 98:024311. https://doi.org/10.1103/ PhysRevB.98.024311
59. Kungl AF, Schmitt S, Klähn J, Müller P, Baumbach A, Dold D, Kugele A, Müller E, Koke C, Kleider M, Mauch C, Breitwieser O, Leng L, Gürtler N, Güttler M, Husmann D, Husmann K, Hartel A, Karasenko V, Grübl A, Schemmel J, Meier K, Petrovici MA (2019) Accelerated physical emulation of bayesian inference in spiking neural networks. Front Neurosci 13:1201. https://www.frontiersin.org/article/10.3389/fnins.2019.01201
60. Lu S, Gao X, Duan L-M (2019) Efficient representation of topologically ordered states with restricted Boltzmann machines. Phys Rev B 99:155136. https://link.aps.org/doi/10.1103/ PhysRevB.99.155136
61. Gao X, Duan L-M (2017) Efficient representation of quantum many-body states with deep neural networks. Nat Commun 8(1):662. https://doi.org/10.1038/s41467-017-00705-2
62. Deng D-L, Li X, Das Sarma S (2017) Quantum entanglement in neural network states. Phys Rev X 7:021021. https://link.aps.org/doi/10.1103/PhysRevX.7.021021
63. Nomura Y, Darmawan AS, Yamaji Y, Imada M (2017) Restricted Boltzmann machine learning for solving strongly correlated quantum systems. Phys Rev B 96:205152. https://link.aps.org/ doi/10.1103/PhysRevB.96.205152
64. Carrasquilla J, Torlai G, Melko RG, Aolita L (2019) Reconstructing quantum states with generative models. Nat Mach Intell 1(3):155–161. https://doi.org/10.1038/s42256-019-0028-1
65. Hartmann MJ, Carleo G (2019) Neural-network approach to dissipative quantum many-body dynamics. Phys Rev Lett 122:250502. https://link.aps.org/doi/10.1103/PhysRevLett. 122.250502
66. Nagy A, Savona V (2019) Variational quantum Monte Carlo method with a neural-network ansatz for open quantum systems. Phys Rev Lett 122:250501. https://link.aps.org/doi/10. 1103/PhysRevLett.122.250501
67. Vicentini F, Biella A, Regnault N, Ciuti C (2019) Variational neural-network ansatz for steady states in open quantum systems. Phys Rev Lett 122:250503. https://link.aps.org/doi/10.1103/ PhysRevLett.122.250503
68. Yoshioka N, Hamazaki R (2019) Constructing neural stationary states for open quantum many-body systems. Phys Rev B 99:214306. https://link.aps.org/doi/10.1103/PhysRevB.99. 214306
69. Torlai G, Mazzola G, Carrasquilla J, Troyer M, Melko R, Carleo G (2018) Neural-network quantum state tomography. Nat Phys 14:447–450. https://doi.org/10.1038/s41567-018-0048-5
70. Torlai G, Melko RG (2018) Latent space purification via neural density operators. Phys Rev Lett 120:240503. https://link.aps.org/doi/10.1103/PhysRevLett.120.240503
71. Torlai G, Timar B, van Nieuwenburg EPL, Levine H, Omran A, Keesling A, Bernien H, Greiner M, Vuletić V, Lukin MD, Melko RG, Endres M (2019) Integrating neural networks with a quantum simulator for state reconstruction. Phys Rev Lett 123:230504. https://link. aps.org/doi/10.1103/PhysRevLett.123.230504

72. Westerhout T, Astrakhantsev N, Tikhonov KS, Katsnelson M, Bagrov AA (2020) General-
    ization properties of neural network approximations to frustrated magnet ground states. Nat
    Commun 11(1):1593. https://doi.org/10.1038/s41467-020-15402-w
73. Saito H (2017) Solving the Bose-Hubbard model with machine learning. J Phys Soc Jpn
    86(9):093001. https://doi.org/10.7566/JPSJ.86.093001
74. Cai Z, Liu J (2018) Approximating quantum many-body wave functions using artificial neural
    networks. Phys Rev B 97:035116. https://link.aps.org/doi/10.1103/PhysRevB.97.035116
75. Deng D-L, Li X, Das Sarma S (2017) Machine learning topological states. Phys Rev B
    96:195145. https://link.aps.org/doi/10.1103/PhysRevB.96.195145
76. Kaubruegger R, Pastori L, Budich JC (2018) Chiral topological phases from artificial neural
    networks. Phys Rev B 97:195136. https://link.aps.org/doi/10.1103/PhysRevB.97.195136
77. Freitas N, Morigi G, Dunjko V (2018) Neural network operations and Susuki-Trotter evo-
    lution of neural network states. Int J Quantum Inf 16(08):1840008. https://doi.org/10.1142/
    S0219749918400087
78. Carleo G, Nomura Y, Imada M (2018) Constructing exact representations of quantum many-
    body systems with deep neural networks. Nat Commun 9(1):5322. https://doi.org/10.1038/
    s41467-018-07520-3
79. Teng P (2018) Machine-learning quantum mechanics: solving quantum mechanics problems
    using radial basis function networks. Phys Rev E 98:033305. https://link.aps.org/doi/10.1103/
    PhysRevE.98.033305
80. Huang Y, Moore JE (2017) Neural network representation of tensor network and chiral states.
    arXiv:1701.06246 [cond-mat.dis-nn]
81. Glasser I, Pancotti N, August M, Rodriguez ID Ignacio Cirac J (2018) Neural-network quan-
    tum states, string-bond states, and chiral topological states. Phys Rev X 8:011006, Jan 2018.
    https://link.aps.org/doi/10.1103/PhysRevX.8.011006
82. Clark SR (2018) Unifying neural-network quantum states and correlator product states via
    tensor networks. J Phys A: Math Theor 51(13):135301. http://stacks.iop.org/1751-8121/51/
    i=13/a=135301
83. Chen J, Cheng S, Xie H, Wang L, Xiang T (2018) Equivalence of restricted Boltzmann
    machines and tensor network states. Phys Rev B 97:085104. https://link.aps.org/doi/10.1103/
    PhysRevB.97.085104
84. Troyer M, Wiese U-J (2005) Computational complexity and fundamental limitations to
    fermionic quantum Monte Carlo simulations. Phys Rev Lett 94:170201. https://link.aps.org/
    doi/10.1103/PhysRevLett.94.170201
85. Anagnostopoulos KN, Nishimura J (2002) New approach to the complex-action problem and
    its application to a nonperturbative study of superstring theory. Phys Rev D 66:106008. https://
    link.aps.org/doi/10.1103/PhysRevD.66.106008
86. Nakamura T, Hatano N, Nishimori H (1992) Reweighting method for quantum Monte Carlo
    simulations with the negative-sign problem. J Phys Soc Jpn 61(10):3494–3502. https://doi.
    org/10.1143/JPSJ.61.3494
87. Loh EY, Gubernatis JE, Scalettar RT, White SR, Scalapino DJ, Sugar RL (1990) Sign problem
    in the numerical simulation of many-electron systems. Phys Rev B 41:9301–9307. https://
    link.aps.org/doi/10.1103/PhysRevB.41.9301
88. Broecker P, Carrasquilla J, Melko RG, Trebst S (2017) Machine learning quantum phases of
    matter beyond the fermion sign problem. Sci Rep 7:8823. https://doi.org/10.1038/s41598-
    017-09098-0
89. Torlai G, Carrasquilla J, Fishman MT, Melko RG, Fisher MPA (2019) Wavefunction posi-
    tivization via automatic differentiation. arXiv:1906.04654 [quant-ph]
90. Hangleiter D, Roth I, Nagaj D, Eisert J (2019) Easing the Monte Carlo sign problem.
    arXiv:1906.02309 [quant-ph]
91. Czischek S, Pawlowski JM, Gasenzer T, Gärttner M (2019) Sampling scheme for neuromor-
    phic simulation of entangled quantum systems. Phys Rev B 100:195120. https://link.aps.org/
    doi/10.1103/PhysRevB.100.195120

92. Sachdev S (2011) Quantum phase transitions, 2nd edn. Cambridge University Press, Cambridge. https://doi.org/10.1017/CBO9780511973765
93. Pfeuty P (1970) The one-dimensional using model with a transverse field. Ann Phys (NY) 57:79–90. https://doi.org/10.1016/0003-4916(70)90270-8
94. Lieb E, Schultz T, Mattis D (1961) Two soluble models of an antiferromagnetic chain. Ann Phys 16(3):407–466. http://www.sciencedirect.com/science/article/pii/0003491661901154
95. Calabrese P, Essler FHL, Fagotti M (2012) Quantum quench in the transverse field using chain: I. time evolution of order parameter correlators. J Stat Mech: Theory Exp 2012(07):P07016. https://doi.org/10.1088%2F1742-5468%2F2012%2F07%2Fp07016
96. Calabrese P, Essler FHL, Fagotti M (2012) Quantum quenches in the transverse field using chain: II. stationary state properties. J Stat Mech: Theory Exp 2012(07):P07022. https://doi.org/10.1088%2F1742-5468%2F2012%2F07%2Fp07022
97. Chiocchetta A, Gambassi A, Diehl S, Marino J (2017) Dynamical crossovers in prethermal critical states. Phys Rev Lett 118:135701. https://link.aps.org/doi/10.1103/PhysRevLett.118.135701
98. Delfino G (2014) Quantum quenches with integrable pre-quench dynamics. J Phys A: Math Theor 47(40):402001. http://stacks.iop.org/1751-8121/47/i=40/a=402001
99. Delfino G, Viti J (2017) On the theory of quantum quenches in near-critical systems. J Phys A: Math Theor 50(8):084004. http://stacks.iop.org/1751-8121/50/i=8/a=084004
100. Karl M, Cakir H, Halimeh JC, Oberthaler MK, Kastner M, Gasenzer T (2017) Universal equilibrium scaling functions at short times after a quench. Phys Rev E 96:022110 Aug. https://link.aps.org/doi/10.1103/PhysRevE.96.022110
101. Di Ventra M, Traversa FL (2018) Perspective: Memcomputing: Leveraging memory and physics to compute efficiently. J Appl Phys 123(18):180901. https://doi.org/10.1063/1.5026506
102. Petrovici MA (2016) Form versus function: theory and models for neuronal substrates. Springer International Publishing, Berlin. https://doi.org/10.1007/978-3-319-39552-4
103. Petrovici MA, Bill J, Bytschok I, Schemmel J, Meier K (2016) Stochastic inference with spiking neurons in the high-conductance state. Phys Rev E 94:042312 Oct. https://link.aps.org/doi/10.1103/PhysRevE.94.042312
104. Schemmel J, Brüderle D, Grübl A, Hock M, Meier K, Millner S (2010) A wafer-scale neuromorphic hardware system for large-scale neural modeling, pp 1947–1950. https://ieeexplore.ieee.org/document/5536970/
105. Buesing L, Bill J, Nessler B, Maass W (2011) Neural dynamics as sampling: a model for stochastic computation in recurrent networks of spiking neurons. PLoS Comput Biol 7(11):1–22. https://doi.org/10.1371/journal.pcbi.1002211
106. Wunderlich T, Kungl AF, Müller E, Hartel A, Stradmann Y, Aamir SA, Grübl A, Heimbrecht A, Schreiber K, Stöckel D, Pehle C, Billaudelle S, Kiene G, Mauch C, Schemmel J, Meier K, Petrovici MA. Demonstrating advantages of neuromorphic computation: a pilot study. Front Neurosci 13:260. https://www.frontiersin.org/article/10.3389/fnins.2019.00260
107. John Stewart Bell (1964) On the Einstein Podolsky Rosen paradox. Physics 1(3):195–200. https://cds.cern.ch/record/111654
108. Bell JS (1966) On the problem of hidden variables in quantum mechanics. Rev Mod Phys 38:447–452. https://link.aps.org/doi/10.1103/RevModPhys.38.447
109. Bell JS, Aspect A (2004) Speakable and unspeakable in quantum mechanics: collected papers on quantum philosophy, 2 edn. Cambridge University Press, Cambridge. https://doi.org/10.1017/CBO9780511815676
110. Greenberger DM, Horne MA, Zeilinger A (1989) Going beyond bell's theorem, pp 69–72. Springer Netherlands, Dordrecht. https://doi.org/10.1007/978-94-017-0849-4_10

# Part I
# Background

# Chapter 2
# Quantum Mechanics and Spin Systems

With the basic content of this thesis being simulations of quantum spin-1/2 systems, we start introducing the underlying concepts and properties of those and of quantum mechanics in general. Clarifying the basics, we provide a starting point to perform approximate simulations of these computationally unfeasible problems. An entire summary of the rudiments of quantum mechanics is given in [1–3], on which Sect. 2.1 is based. Section 2.2 builds on [1, 3, 4], where the properties of quantum spin systems are discussed in detail. To introduce quantum entanglement, Sect. 2.3 is based on a detailed summary in [1, 3].

Furthermore, we introduce in this chapter the models considered later in this thesis for benchmarking simulation methods. In particular these are the Bell and Greenberger–Horne–Zeilinger states, as well as the transverse-field Ising model with and without an additional longitudinal field. We also introduce the commonly used simulation method exact diagonalization and the approximate time-dependent density-matrix renormalization group method. Both are well understood and suitable for benchmarking new approximation schemes.

## 2.1 Concepts of Quantum Mechanics

In quantum mechanics we consider systems of particles with different properties being represented as vectors of a complex Hilbert space $\mathcal{H}$. This is a complete vector space under the norm induced by its scalar product. Thus, a pure state of an $N$-particle quantum system can be described via a vector $|\Psi\rangle = (\Psi_1, \ldots, \Psi_N)^{\mathrm{T}}$ which is called the state vector and is an element of the Hilbert space $\mathcal{H}$. We here use the Dirac (bra-ket) notation, where the "ket" $|\Psi\rangle$ denotes a column vector and the "bra" $\langle\Psi|$

© The Editor(s) (if applicable) and The Author(s), under exclusive
license to Springer Nature Switzerland AG 2020
S. Czischek, *Neural-Network Simulation of Strongly Correlated Quantum Systems*,
Springer Theses, https://doi.org/10.1007/978-3-030-52715-0_2

is its conjugate transpose. It is hence the row vector $\langle \Psi | = (\Psi_1^*, \ldots, \Psi_N^*)$ with the star denoting complex conjugation.

The state vectors of quantum systems can be chosen normalized,

$$\| |\Psi\rangle \| = \sqrt{\langle \Psi | \Psi \rangle}$$
$$= 1, \tag{2.1}$$

and they can be expanded in the basis states $|\mathbf{v}_i\rangle$ of the Hilbert space,

$$|\Psi\rangle = \sum_{i=1}^{d_H} c_{\mathbf{v}_i} |\mathbf{v}_i\rangle, \tag{2.2}$$

with complex coefficients $c_{\mathbf{v}_i} \in \mathbb{C}$ and Hilbert space dimension $d_H$. The state vector is thus a superposition of basis states.

Furthermore, we introduce the wave function $\Psi(\mathbf{v})$ as the overlap of the state vector with a basis vector $\langle \mathbf{v} |$,

$$\Psi(\mathbf{v}) = \langle \mathbf{v} | \Psi \rangle, \tag{2.3}$$

which provides a complex-valued amplitude for each basis state. A probability amplitude $P(\mathbf{v})$ for the corresponding basis state is given by the squared wave function,

$$P(\mathbf{v}) = |\Psi(\mathbf{v})|^2, \tag{2.4}$$

emphasizing the choice of normalized wave functions, since then also $P(\mathbf{v})$ describes a normalized probability.

Another possibility to describe the state of a quantum system is to introduce the density matrix as

$$\hat{\rho} = |\Psi\rangle\langle\Psi|, \tag{2.5}$$

with the hat denoting that $\hat{\rho}$ is a matrix. Plugging the expansion of the state vector in terms of the basis states [Eq. (2.1)] into Eq. (2.5) shows that the density matrix has the diagonal entries $\hat{\rho}_{i,i} = |c_{\mathbf{v}_i}|^2$. These describe the probabilities of the system to be in the state $|\mathbf{v}_i\rangle$, so that we get the property

$$\mathrm{Tr}\left[\hat{\rho}\right] = \sum_{i=1}^{d_H} \hat{\rho}_{i,i} = 1, \tag{2.6}$$

stating that all probabilities sum up to one. Additionally, the density matrix is Hermitian by construction and it is positive definite, since all probabilities are positive.

Given the possibility to express a quantum state via the state vector or the density matrix, we can introduce a procedure to perform measurements on the system, which corresponds to calculating observables. Any observable acting on a quantum system can be represented by a self-adjoint linear operator $\hat{O}$ acting on elements of the Hilbert space. As the Hilbert space is a vector space, this operator can be represented by a Hermitian matrix. Here and in the remainder of this chapter we use hats to denote operators, as already done for the density matrix in Eq. (2.5). We will omit the hats again by the end of this chapter, when the basic concepts of quantum mechanics have become clear and it is obvious from the context which quantities are matrices.

When applying such an operator onto a state vector of a quantum system, which we call performing a measurement, the state is transformed into an eigenstate of the operator. It is then represented by a state vector which is an eigenvector of the corresponding matrix with the measurement outcome being the corresponding eigenvalue. Thus, the state of the system is changed when a measurement is performed.

In contrast to classical systems, the outcome of a measurement is not deterministic. If the system is in a superposition of different eigenstates of the operator, each measurement provides exactly one possible outcome. While each measurement can lead to a different result, the outcome of infinitely many measurements is distributed according to the probability distribution underlying the probabilistic superposition of states.

We are hence interested in the average outcome of measuring an observable $\hat{O}$, which can be calculated using the state vector or the density matrix,

$$
\begin{aligned}
\langle \hat{O} \rangle &= \langle \Psi | \hat{O} | \Psi \rangle \\
&= \sum_{i=1}^{d_{\mathrm{H}}} \langle \Psi | \hat{O} | \mathbf{v}_i \rangle \langle \mathbf{v}_i | \Psi \rangle \\
&= \sum_{i=1}^{d_{\mathrm{H}}} \langle \mathbf{v}_i | \Psi \rangle \langle \Psi | \hat{O} | \mathbf{v}_i \rangle \\
&= \mathrm{Tr} \left[ \hat{\rho} \hat{O} \right],
\end{aligned}
\tag{2.7}
$$

with brackets $\langle \hat{O} \rangle$ denoting expectation values. The trace is given by the sum over all basis states,

$$
\mathrm{Tr} \left[ \hat{O} \right] = \sum_{i=1}^{d_{\mathrm{H}}} \langle \mathbf{v}_i | \hat{O} | \mathbf{v}_i \rangle,
\tag{2.8}
$$

and we multiply with one in the second line,

$$
\sum_{i=1}^{d_{\mathrm{H}}} | \mathbf{v}_i \rangle \langle \mathbf{v}_i | = 1.
\tag{2.9}
$$

Due to the fact that after a measurement the state of a system is changed into an eigenstate of the corresponding operator, two observables whose operators do not commute, $\hat{O}_A\hat{O}_B \neq \hat{O}_B\hat{O}_A$, cannot be measured at the same time as they have different eigenbases. This leads to Heisenberg's uncertainty relation, which states that position and momentum of a quantum particle cannot be measured exactly at the same time. They are only accessible within some uncertainty, $\Delta\hat{x}\Delta\hat{p} \geq \hbar/2$ due to $[\hat{x}, \hat{p}] = \hat{x}\hat{p} - \hat{p}\hat{x} = i\hbar$, with the reduced Planck's constant $\hbar = h/(2\pi)$.

Knowing how to apply operators onto a state vector of a quantum system, we can also calculate the system's time evolution under some Hamiltonian operator $\hat{H}$. The corresponding equation of motion is given by the Schrödinger equation,

$$i\hbar\frac{\partial}{\partial t}|\Psi(t)\rangle = \hat{H}|\Psi(t)\rangle. \tag{2.10}$$

For the case of the system being in a stationary state, this equation simplifies to the time-independent Schrödinger equation,

$$\hat{H}|\Psi\rangle = E|\Psi\rangle, \tag{2.11}$$

where $E$ is the energy of the system and an eigenvalue of $\hat{H}$. Equation (2.10) can also be generalized to provide the von-Neumann equation of motion for the density matrix,

$$i\hbar\frac{\partial}{\partial t}\hat{\rho}(t) = \left[\hat{H}, \hat{\rho}(t)\right], \tag{2.12}$$

with the commutator $[\hat{A}, \hat{B}] = \hat{A}\hat{B} - \hat{B}\hat{A}$.

For a quantum system where each single particle lives in a Hilbert space of dimension $d_H = d$, the total Hilbert space $\mathcal{H}$ of a system with $N$ distinguishable such particles is the tensor product of all individual Hilbert spaces $\mathcal{H}^{(n)}$,

$$\begin{aligned}\mathcal{H} = \mathcal{H}^{\otimes N} &= \bigotimes_{n=1}^{N} \mathcal{H}^{(n)}\\ &= \mathcal{H}^{(1)} \otimes \mathcal{H}^{(2)} \otimes \cdots \otimes \mathcal{H}^{(N)}.\end{aligned} \tag{2.13}$$

Thus, the total Hilbert space has dimension $d_H = d^N$, where $d$ is referred to as the local dimension of a single particle. Hence, $d^N$ basis states are necessary to fully describe the total Hilbert space. This demonstrates the so-called curse of dimensionality in quantum mechanics, which states the problem of the exponentially growing Hilbert space, where superpositions of all states can appear.

If we instead consider indistinguishable particles, exchanging two of them must not change the system state. Thus, the wave functions of the states with two interchanged particles can at most differ by an overall phase, as this does not change the probability amplitude. So, if we consider two particles in states $|\mathbf{v}_1\rangle$ and $|\mathbf{v}_2\rangle$, we get

$$\Psi\left(\mathbf{v}_1, \mathbf{v}_2\right) = \left(\langle \mathbf{v}_1| \otimes \langle \mathbf{v}_2|\right)|\Psi\rangle$$
$$= \langle \mathbf{v}_1, \mathbf{v}_2| \Psi\rangle \qquad (2.14)$$
$$= \exp\left[i\varphi\right]\Psi\left(\mathbf{v}_2, \mathbf{v}_1\right),$$

with a global phase $\exp[i\varphi]$. In three spatial dimensions, exchanging two particles twice yields

$$\Psi\left(\mathbf{v}_1, \mathbf{v}_2\right) = \exp\left[2i\varphi\right]\Psi\left(\mathbf{v}_1, \mathbf{v}_2\right) \qquad (2.15)$$
$$\Rightarrow \exp\left[2i\varphi\right] = \pm 1 \qquad (2.16)$$
$$\Rightarrow \Psi\left(\mathbf{v}_2, \mathbf{v}_1\right) = \pm \Psi\left(\mathbf{v}_1, \mathbf{v}_2\right). \qquad (2.17)$$

Here we also experience the case that we have to exchange the particles four times to get back the original wave function without an overall phase. We can divide the space of states into two orthogonal sub-spaces, one containing the states where exchanging two particles yields a minus sign (anti-symmetric under permutations) and one where it does not (symmetric under permutations). Particles in the first subspace are called fermions, while particles in the second subspace are called bosons. Since the wave function of fermions is anti-symmetric under permutations, two such particles cannot be in the same state, as

$$\Psi\left(\mathbf{v}_1, \mathbf{v}_1\right) = -\Psi\left(\mathbf{v}_1, \mathbf{v}_1\right) \qquad (2.18)$$
$$\Rightarrow \Psi\left(\mathbf{v}_1, \mathbf{v}_1\right) = 0. \qquad (2.19)$$

Thus, the probability for such a configuration is zero.

A quantum many-body system with an unknown number of particles can be described in the so-called Fock space $\mathcal{F}$, which is the product state of the $n$-particle Hilbert spaces for all possible $n$,

$$\mathcal{F} = \bigoplus_{n=0}^{\infty} \hat{S}_{\pm}\mathcal{H}^{\otimes n}. \qquad (2.20)$$

$\hat{S}_{\pm}$ denote the symmetrization or anti-symmetrization operators used for bosonic or fermionic systems, respectively. A basis of the Fock space is given by the so-called occupation-number-basis, which specifies how many particles occupy a single-site wave function. The basis states are hence characterized by these occupation numbers. For bosonic systems the particle number can generally take any integer value between zero and infinity, while for fermionic systems the occupation number is restricted to either zero or one, as no two fermions can be in the same state.

Using the Fock space representation, one can define creation operators $\hat{a}_i^{\dagger}$, which create a particle in state $i$. Their conjugate transpose, $\hat{a}_i$, are the annihilation operators which annihilate a particle in state $i$. Thus, any state in the occupation-number-basis can be expressed in terms of creation operators,

$$|n_1, \ldots, n_K\rangle = \prod_{i=1}^{K} \left(\hat{a}_i^\dagger\right)^{n_i} |0\rangle. \tag{2.21}$$

Here we consider a system with $K$ possible states each individual particle can be in and the values $n_1, \ldots, n_K$ are the occupation numbers of the individual states. Thus, $|0\rangle = |0, \ldots, 0\rangle$ is the vacuum state with all single-site wave functions being unoccupied. The eigenstate $|\alpha\rangle$ of the annihilation operator,

$$\hat{a}^\dagger |\alpha\rangle = \alpha |\alpha\rangle, \tag{2.22}$$

is called a coherent state.

These creation and annihilation operators look differently for bosonic and fermionic systems and they even have different properties. They all satisfy canonical commutation relations, which for bosonic systems are given by

$$\begin{aligned}
\left[\hat{a}_i, \hat{a}_j\right] &= \left[\hat{a}_i^\dagger, \hat{a}_j^\dagger\right] = 0, \\
\left[\hat{a}_i, \hat{a}_j^\dagger\right] &= -\left[\hat{a}_i^\dagger, \hat{a}_j\right] = \delta_{i,j},
\end{aligned} \tag{2.23}$$

and for fermionic systems they read

$$\begin{aligned}
\left\{\hat{a}_i^\dagger, \hat{a}_j\right\} &= \left\{\hat{a}_i, \hat{a}_j^\dagger\right\} = \delta_{i,j}, \\
\left\{\hat{a}_i, \hat{a}_j\right\} &= \left\{\hat{a}_i^\dagger, \hat{a}_j^\dagger\right\} = 0,
\end{aligned} \tag{2.24}$$

with the curly brackets denoting the anti-commutator $\{\hat{A}, \hat{B}\} = \hat{A}\hat{B} + \hat{B}\hat{A}$. Thus, when expressing a state in the occupation-number-basis in terms of the creation operators as stated in Eq. (2.21), it is important for fermionic systems in which order the operators are applied, as interchanging them yields a minus sign. The order can be chosen freely once, but it must be kept consistent during further calculations.

## 2.2  Properties of Spin Systems

Considering elementary particles, such as electrons, Otto Stern and Walter Gerlach found in 1922 that they have a discrete intrinsic angular momentum [2, 5]. Nowadays this experiment is known as the Stern–Gerlach experiment. The intrinsic angular momentum, referred to as the spin of a particle, is closely related to the angular momentum of a classical spinning object, except for it being discretized.

Therefore, it can be described via spin operators $\hat{\mathbf{S}}$, which show the same properties as angular momentum operators, i.e., they satisfy

$$\left[ \hat{S}^{\alpha}, \hat{S}^{\beta} \right] = i\hbar \hat{S}^{\gamma} \varepsilon_{\alpha\beta\gamma}, \tag{2.25}$$

with $\alpha, \beta, \gamma \in \{x, y, z\}$ and the fully anti-symmetric Levi-Civita tensor $\varepsilon_{\alpha\beta\gamma}$. The sum over repeated indices is implied. Furthermore, the spin operators satisfy

$$\left[ \hat{S}^{z}, \hat{\mathbf{S}}^{2} \right] = 0, \tag{2.26}$$

for $\hat{\mathbf{S}}^{2} = (\hat{S}^{x})^{2} + (\hat{S}^{y})^{2} + (\hat{S}^{z})^{2}$.

Considering a single particle, its state is uniquely defined in terms of quantum numbers. In order to fully describe the spin state of a particle, we use the spin quantum number $l$, which is an integer or a half integer, and the spin projection quantum number $s$, whose value runs from $-l$ to $l$ in steps of one. Thus, we can express a spin state as $|l, s\rangle$, where $l$ is a fixed property of the considered particle, for an electron $l = 1/2$.

If we express the quantum system in the eigenbasis of the spin operator $\hat{S}^{z}$, applying it to a spin state yields the spin projection quantum number,

$$\hat{S}^{z}|l, s\rangle = \hbar s|l, s\rangle. \tag{2.27}$$

On the other hand, the spin quantum number can be obtained by applying $\hat{\mathbf{S}}^{2}$,

$$\hat{\mathbf{S}}^{2}|l, s\rangle = \hbar^{2}l\,(l+1)\,|l, s\rangle. \tag{2.28}$$

Here we choose without loss of generality the spin states as basis states of $\hat{S}^{z}$. The operators $\hat{S}^{x}$ and $\hat{S}^{y}$ do not commute with $\hat{S}^{z}$ and hence have different eigenbases.

In the remainder of this thesis we consider spin-1/2 systems, so we discuss those in more detail. For a spin-1/2 particle, such as an electron, the quantum numbers are $l = 1/2, s = \pm 1/2$. Hence, only two possible spin states exist,

$$|l = 1/2, s = 1/2\rangle =: |\uparrow\rangle \equiv |1\rangle, \tag{2.29}$$
$$|l = 1/2, s = -1/2\rangle =: |\downarrow\rangle \equiv |-1\rangle. \tag{2.30}$$

These are commonly referred to as the spin-up and spin-down states, so we introduce the notation with arrows pointing up or down. Another common notation is to enumerate the states by $|-1\rangle$ and $|1\rangle$.

The two basis states are defined to be normalized and orthogonal,

$$\begin{aligned} \langle\uparrow|\uparrow\rangle = \langle\downarrow|\downarrow\rangle = 1, \\ \langle\uparrow|\downarrow\rangle = \langle\downarrow|\uparrow\rangle = 0, \end{aligned} \tag{2.31}$$

and we consider the spin in the basis of the $\hat{S}^{z}$ operator, so that

$$\hat{S}^z|\uparrow\rangle = \frac{\hbar}{2}|\uparrow\rangle,$$

$$\hat{S}^z|\downarrow\rangle = -\frac{\hbar}{2}|\downarrow\rangle. \tag{2.32}$$

Analogously to the creation and annihilation operators in the Fock space representation we can define raising and lowering operators $\hat{S}^\pm$ for the spin states,

$$\hat{S}^+|\uparrow\rangle = \hat{S}^-|\downarrow\rangle = 0,$$

$$\hat{S}^+|\downarrow\rangle = |\uparrow\rangle, \tag{2.33}$$

$$\hat{S}^-|\uparrow\rangle = |\downarrow\rangle.$$

Any spin-1/2 particle can be in an arbitrary superposition of the two basis states,

$$|\Psi\rangle = \alpha|\uparrow\rangle + \beta|\downarrow\rangle,$$

$$|\alpha|^2 + |\beta|^2 = 1, \tag{2.34}$$

where the condition in the second line ensures that the state vector is normalized.

To make the quantum spin state description more concrete, we can define the basis states in a two-dimensional Hilbert space $\mathcal{H} = \mathbb{C}^2$, where a common choice is

$$|\uparrow\rangle = \begin{pmatrix} 1 \\ 0 \end{pmatrix}, \tag{2.35}$$

$$|\downarrow\rangle = \begin{pmatrix} 0 \\ 1 \end{pmatrix}. \tag{2.36}$$

It can be shown that for this choice the spin operators can be defined via the Pauli matrices $\hat{\sigma}^x, \hat{\sigma}^y, \hat{\sigma}^z$ as

$$\hat{S}^x = \frac{\hbar}{2}\hat{\sigma}^x = \frac{\hbar}{2}\begin{pmatrix} 0 & 1 \\ 1 & 0 \end{pmatrix}, \tag{2.37}$$

$$\hat{S}^y = \frac{\hbar}{2}\hat{\sigma}^y = \frac{\hbar}{2}\begin{pmatrix} 0 & -i \\ i & 0 \end{pmatrix}, \tag{2.38}$$

$$\hat{S}^z = \frac{\hbar}{2}\hat{\sigma}^z = \frac{\hbar}{2}\begin{pmatrix} 1 & 0 \\ 0 & -1 \end{pmatrix}. \tag{2.39}$$

These fulfill the properties in Eqs. (2.25)–(2.28), as can be checked straightforwardly.

In the remainder of this thesis we set $\hbar = 1$ if not stated otherwise, fixing the units. We furthermore switch to the representation of spin-1/2 systems using the spin values $\pm 1$ instead of $\pm 1/2$, so that we can directly use the Pauli matrices as spin operators, $\hat{S}^\alpha = \hat{\sigma}^\alpha, \alpha \in \{x, y, z\}$. This is possible since the eigenvalues of $\hat{\sigma}^z$ are $\pm 1$, so that $\hat{\sigma}^z|\uparrow\rangle = |\uparrow\rangle$ and $\hat{\sigma}^z|\downarrow\rangle = -|\downarrow\rangle$. This step can be done without loss of generality but needs to be kept consistent. We can then introduce raising and lowering operators as

$$\hat{S}^{\pm} = \frac{1}{2}\left(\hat{S}^x \pm i\hat{S}^y\right). \tag{2.40}$$

The time evolution of a single spin state under a Hamiltonian $\hat{H}$ can be calculated using unitary operators $\hat{U}$,

$$\hat{U} = \exp\left[-it\hat{H}\right], \tag{2.41}$$

with $\hat{U}\hat{U}^{\dagger} = \hat{\mathbb{1}}$. Here $\hat{U}^{\dagger}$ denotes the conjugate transpose (adjoint) and $\hat{\mathbb{1}}$ is the identity matrix. The Hamiltonian operator acting on a single spin state can moreover be expressed in terms of the Pauli spin operators,

$$\hat{H} = \sum_{\substack{\alpha \in \\ \{0,x,y,z\}}} n_{\alpha}\hat{\sigma}^{\alpha}, \tag{2.42}$$

with some unit vector $\mathbf{n}$ and identity operator $\hat{\sigma}^0 = \hat{\mathbb{1}}$. In this case the unitary operators $\hat{U}$ form a group SU(2), with the generators given by the spin operators, and provide a solution to the Schrödinger equation, Eq. (2.10).

If larger systems consisting of many spins are considered, the Hilbert space of the full system is the tensor product of the Hilbert spaces of the individual particles. Thus, these systems can be treated the way introduced in Sect. 2.1. The Hilbert space dimension of a system of $N$ spin-1/2 particles is $2^N$, so it scales exponentially with the system size. This makes computations in these systems extremely expensive and unfeasible for large system sizes, even though particles with the smallest individual Hilbert space dimension are considered.

Measurements can in general also be performed only on a part of the spins in the system, so we can measure the spin values at sites $i$ and $j$ in the chain via the operator

$$\hat{S}_i^z \hat{S}_j^z \equiv \hat{\mathbb{1}} \otimes \hat{\mathbb{1}} \otimes \cdots \otimes \hat{\sigma}_i^z \otimes \cdots \otimes \hat{\sigma}_j^z \otimes \cdots \otimes \hat{\mathbb{1}}, \tag{2.43}$$

with identity matrices $\hat{\mathbb{1}}$ applied to all remaining spins. This we abbreviate as $\hat{S}_i^z \hat{S}_j^z = \hat{\sigma}_i^z \hat{\sigma}_j^z$, leaving out the identity matrices for convenience sake.

A common way to visualize a single spin-1/2 state is via the so-called Bloch sphere. Since a general state vector of a single spin-1/2 particle,

$$|\Psi\rangle = \alpha|\uparrow\rangle + \beta|\downarrow\rangle, \tag{2.44}$$

is normalized, we get the restriction stated in Eq. (2.34). Furthermore, the state vector is invariant under a global phase as this does not influence the probabilities $|\alpha|^2$ and $|\beta|^2$. Expressing the complex parameters $\alpha$, $\beta$ in terms of real amplitudes $r_{\alpha}$, $r_{\beta}$ and phases $\varphi_\alpha$, $\varphi_\beta$, we get

$$
\begin{aligned}
|\Psi\rangle &= r_\alpha \exp[i\varphi_\alpha]\,|\!\uparrow\rangle + r_\beta \exp[i\varphi_\beta]\,|\!\downarrow\rangle \\
&= \exp[i\varphi_\alpha]\left\{ r_\alpha|\!\uparrow\rangle + r_\beta \exp[i\varphi_\beta - i\varphi_\alpha]\,|\!\downarrow\rangle \right\} \\
&= r_\alpha|\!\uparrow\rangle + r_\beta \exp\left[i\left(\varphi_\beta - \varphi_\alpha\right)\right]|\!\downarrow\rangle \\
&=: r_\alpha|\!\uparrow\rangle + r_\beta \exp[i\varphi]\,|\!\downarrow\rangle,
\end{aligned}
\tag{2.45}
$$

where we use the invariance under a global phase when going from the second to the third line.

The normalization constraint, Eq. (2.34), then corresponds to the equation of a unit sphere in cartesian coordinates. We can hence express Eq. (2.45) in spherical coordinates with azimuthal angle $\theta$ and polar angle $\phi$,

$$
|\Psi\rangle = \cos\left[\frac{\theta}{2}\right]|\!\uparrow\rangle + \exp[i\phi]\sin\left[\frac{\theta}{2}\right]|\!\downarrow\rangle.
\tag{2.46}
$$

The half-angles result from the fact that opposite states in the upper and lower hemisphere differ by a phase factor of $-1$, so that points on the upper hemisphere can be mapped onto the whole sphere.

With these two angles $\theta$ and $\phi$, any single spin state can be represented as one point on the surface of a sphere, defined as the Bloch sphere, as depicted in Fig. 2.1. The north pole of the Bloch sphere corresponds to the state $|\!\uparrow\rangle$ and the south pole

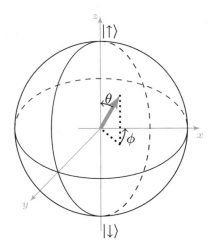

**Fig. 2.1** Visualization of the state vector of a single spin-1/2 particle on a Bloch sphere according to Eq. (2.46). The spin state is fully described by the two angles $\theta$ and $\phi$, which are defined as depicted. The arrow denotes a possible state vector. Any spin state can be represented by a point on the sphere, where all points must lie on the surface due to the normalization of the state vectors [Eq. (2.34)]. The north pole of the Bloch sphere corresponds to the state $|\!\uparrow\rangle$, while the south pole corresponds to $|\!\downarrow\rangle$

corresponds to $|\downarrow\rangle$, while any other point on the surface is defined by the angles $\theta$ and $\phi$. The spin can only live on the surface of the sphere due to the normalization condition, Eq. (2.34). There is, however, no general method to visualize a system of $N$ spins. Considering a system in a separable state, i.e., an $N$-spin state that can be expressed as a tensor product of the individual spin states, each spin can be represented by an individual Bloch sphere. Nevertheless, there is no ansatz to include (quantum) correlations into the visualization. Thus, the Bloch sphere picture is basically used to only visualize single spin particles.

## 2.3 Quantum Entanglement

So far, we have considered quantum systems whose states can be described by a single state vector, which are referred to as pure states. However, most times in nature quantum systems are not in pure states, but in probabilistic mixtures of pure states, so-called mixed states. These can be described by a density matrix which is the sum of many pure density matrices. The contribution of each individual matrix to the mixture is weighted by the probability $p_i$, with which the system is in the corresponding state,

$$
\begin{aligned}
\hat{\rho}^{\text{mixed}} &= \sum_i p_i \hat{\rho}_i^{\text{pure}} \\
&= \sum_i p_i |\Psi_i\rangle\langle\Psi_i|.
\end{aligned}
\tag{2.47}
$$

Here the sum runs over all contributing pure density matrices and the index $i$ denotes the individual pure contributions. This state can then not be expressed by a single state vector anymore.

It can even be measured how pure a quantum state is. To derive a measure for the purity, we consider the squared density matrix,

$$
\begin{aligned}
\left(\hat{\rho}^{\text{mixed}}\right)^2 &= \sum_{i,j} p_i p_j \hat{\rho}_i^{\text{pure}} \hat{\rho}_j^{\text{pure}} \\
&= \sum_{i,j} p_i p_j |\Psi_i\rangle\langle\Psi_i |\Psi_j\rangle\langle\Psi_j| \\
&= \sum_{i,j} p_i p_j |\Psi_i\rangle\langle\Psi_j| \delta_{i,j} \\
&= \sum_i p_i^2 \hat{\rho}_i^{\text{pure}},
\end{aligned}
\tag{2.48}
$$

using the orthogonality of the basis states and their normalization. Taking into account that for a pure state there exists one pure density matrix $\hat{\rho}_i^{\text{pure}}$ in which the system is found with probability $p_i = 1$, while for all other $j \neq i$ we have $p_j = 0$, it follows,

$$\left(\hat{\rho}^{\mathrm{pure}}\right)^2 = \hat{\rho}^{\mathrm{pure}}, \tag{2.49}$$

$$\mathrm{Tr}\left[\left(\hat{\rho}^{\mathrm{pure}}\right)^2\right] = \mathrm{Tr}\left[\hat{\rho}^{\mathrm{pure}}\right] = 1. \tag{2.50}$$

For mixed states we find,

$$\mathrm{Tr}\left[\left(\hat{\rho}^{\mathrm{mixed}}\right)^2\right] = \sum_i p_i^2 \mathrm{Tr}\left[\hat{\rho}_i^{\mathrm{pure}}\right]$$

$$= \sum_i p_i^2 \tag{2.51}$$

$$< 1.$$

Thus, $\mathrm{Tr}[\hat{\rho}^2]$ can be used as a measure for the purity of a state. A maximally mixed state hence minimizes this purity. This is the case if the density matrix is an identity matrix normalized by the Hilbert space dimension $d_{\mathrm{H}} = d^N$ with local dimension $d$ of a single particle ($d = 2$ for a spin-1/2 particle),

$$\hat{\rho} = \frac{\hat{\mathbb{1}}}{d^N} \tag{2.52}$$

$$\Rightarrow \mathrm{Tr}\left[\hat{\rho}^2\right] = \frac{1}{d^N}, \tag{2.53}$$

which is the smallest possible value of the purity.

Another phenomenon found in quantum systems is so-called quantum entanglement, which led to a lot of confusion since it has been experienced by Einstein, Podolsky and Rosen [6]. Here we introduce entanglement and show how to measure it, while details and properties are discussed in Sect. 2.4.1. A general pure state $|\Psi\rangle$ of a multi-particle system is called separable if it can be written as the tensor product of state vectors of two subsystems $A$ and $B$ resulting from bi-partitioning the system,

$$|\Psi\rangle = |\Psi_A\rangle \otimes |\Psi_B\rangle. \tag{2.54}$$

If this is not possible, the state is called entangled.

If we look at a system consisting of two spin-1/2 particles, they can be in the pure state

$$|\Psi\rangle = \frac{1}{\sqrt{2}}\left[|\downarrow\downarrow\rangle + |\uparrow\uparrow\rangle\right]. \tag{2.55}$$

The state vector can be expressed in terms of the basis states as

$$\begin{aligned}
|\Psi\rangle &= |\Psi_1\rangle \otimes |\Psi_2\rangle \\
&= (\alpha_1|\downarrow\rangle + \beta_1|\uparrow\rangle)) \otimes (\alpha_2|\downarrow\rangle + \beta_2|\uparrow\rangle)) \\
&= \alpha_1\alpha_2|\downarrow\downarrow\rangle + \alpha_1\beta_2|\downarrow\uparrow\rangle + \beta_1\alpha_2|\uparrow\downarrow\rangle + \beta_1\beta_2|\uparrow\uparrow\rangle,
\end{aligned} \tag{2.56}$$

but there is no possibility to find values for the coefficients $\alpha_1, \beta_1, \alpha_2, \beta_2$ such that $\alpha_1\alpha_2 = \beta_1\beta_2 = 1/\sqrt{2}$ and $\alpha_1\beta_2 = \beta_1\alpha_2 = 0$. Hence, this state is not separable, and we call it entangled.

To generally see if a system is entangled or not, we introduce the reduced density matrix $\hat{\rho}_A$ of the subsystem $A$, which can be calculated from the total density matrix by tracing over the subsystem $B$,

$$\hat{\rho}_A = \mathrm{Tr}_B\left[\hat{\rho}\right]. \tag{2.57}$$

The reduced density matrix of a separable pure state is again pure, so that we find entanglement if

$$\mathrm{Tr}\left[\hat{\rho}_A^2\right] < 1. \tag{2.58}$$

A measure for the amount of entanglement is given by the entanglement entropy, which is the entropy of the subsystem. In quantum mechanics different definitions for entropy exist. A common choice to consider is the Rényi entropy of order $n$, where $n > 1$ is an integer,

$$S_n\left(\rho\right) = \frac{1}{1-n}\ln\left[\mathrm{Tr}\left(\hat{\rho}^n\right)\right]. \tag{2.59}$$

For $n \to 1$, this expression converges to the von-Neumann entropy,

$$S_{\mathrm{vN}}\left(\hat{\rho}\right) = -\mathrm{Tr}\left[\hat{\rho}\ln\left(\hat{\rho}\right)\right]. \tag{2.60}$$

Calculating the entropy of a subsystem $A$, no matter which definition of entropy is used, yields zero if the system is separable. Moreover, it grows with increasing entanglement in the system. If the entanglement entropy grows proportional to the boundary between the two subsystems, the system is said to have area-law entanglement. On the other hand, it is said to have volume-law entanglement if the entropy grows proportional to the volume of the subsystem $A$.

A maximally mixed state with $\hat{\rho} = \hat{\mathbb{1}}/d^N$ has von-Neumann entanglement entropy

$$\begin{aligned}
S_{\mathrm{vN}}\left(\hat{\rho}_A\right) &= -\mathrm{Tr}\left[\frac{\hat{\mathbb{1}}}{d^N}\ln\left(\frac{\hat{\mathbb{1}}}{d^N}\right)\right] \\
&= N\ln\left[d\right],
\end{aligned} \tag{2.61}$$

which is the maximum value that can be reached.

## 2.4   Entangled Spin-1/2 States

### 2.4.1   Bell State and Bell's Inequality

Having introduced quantum spin-1/2 systems and quantum entanglement, we study examples of small entangled spin systems. These serve as toy-models for analyzing entanglement as they can be solved analytically and are hence fully understood. The smallest of such systems is the Bell state, commonly also referred to as Bell pair, which consists of $N = 2$ spin-1/2 particles with state vector

$$|\Psi^{BP}\rangle := \frac{1}{\sqrt{2}} \left( |\uparrow\downarrow\rangle + |\downarrow\uparrow\rangle \right). \tag{2.62}$$

This spin pair is highly entangled as can already be seen from the state vector since, when performing a measurement on one particle and finding it in a specific state, the state of the other particle is fixed [7–9].

This is a property that has been discussed intensively by J. S. Bell in coherence with the Einstein–Podolsky–Rosen (EPR) paradox. In this gedanken experiment the entanglement between two particles is used to determine position or momentum of one particle with certainty by only measuring the corresponding observable on the other particle. While the EPR paradox could not be explained by the existing quantum mechanics at that time, Einstein, Podolsky and Rosen concluded that the theory is incomplete as physical reality needs a theoretical counterpart in any situation [6]. The theory at this time was based on locality and realism, together often known as "local hidden-variable theory". Here locality refers to the idea that particles cannot be entangled over causally disconnected distances since the correlations are assumed to be encoded by means of locally-acting hidden variables. (Local) realism, on the other hand, refers to the assumption that for any "particle", there is a pre-existing value for the outcome of any kind of measurement.

While there were many debates about the theory of local hidden variables at those times, Bell showed that quantum entanglement cannot be caused by such local hidden variables. This he did by finding a quantity for a system consisting of two spin-1/2 particles that has an upper limit when measured in a classical system, while this limit is violated when considering systems with quantum entanglement [7–9].

While Bell's inequality in its original form is restricted to specific cases and not straightforwardly realizable in experiments, Clauser, Horne, Shimony and Holt introduced a more general and directly measurable form of it, the CHSH-inequality [10, 11]. The quantity considered in the CHSH-inequality is

$$\mathcal{B} := \langle \hat{A}_1 \hat{B}_1 \rangle + \langle \hat{A}_1 \hat{B}_2 \rangle + \langle \hat{A}_2 \hat{B}_1 \rangle - \langle \hat{A}_2 \hat{B}_2 \rangle. \tag{2.63}$$

Here $\hat{A}_1$, $\hat{A}_2$ are two possible observables whose measurements can be performed on the first particle and $\hat{B}_1$, $\hat{B}_2$ are two observables whose measurements can be performed on the second particle. The system is prepared in an initial state and causally separated measurements are performed on the two particles [10, 11].

As this quantity is defined for spin-1/2 particles, all measurements have two possible outcomes, namely $\pm 1$. This implies $|\langle \hat{A}_i \rangle| \leq 1$, $|\langle \hat{B}_i \rangle| \leq 1$ for $i \in \{1, 2\}$ using the triangular inequality. Applying this inequality once more, we find

$$
\begin{aligned}
|\mathcal{B}| &= \left| \langle \hat{A}_1 \hat{B}_1 \rangle + \langle \hat{A}_1 \hat{B}_2 \rangle + \langle \hat{A}_2 \hat{B}_1 \rangle - \langle \hat{A}_2 \hat{B}_2 \rangle \right| \\
&\leq \left| \langle \hat{A}_1 \hat{B}_1 \rangle \right| + \left| \langle \hat{A}_1 \hat{B}_2 \rangle \right| + \left| \langle \hat{A}_2 \hat{B}_1 \rangle \right| - \left| \langle \hat{A}_2 \hat{B}_2 \rangle \right| \quad (2.64) \\
&\leq 2
\end{aligned}
$$

$$
\Rightarrow |\mathcal{B}| \leq 2. \quad (2.65)
$$

This inequality provides an upper limit for the CHSH-observable $\mathcal{B}$ and is referred to as the CHSH inequality.

We can now take a look at a direct quantum mechanical example and choose

$$
\begin{aligned}
\hat{A}_1 &= \hat{\sigma}_1^x, \\
\hat{A}_2 &= \hat{\sigma}_1^z, \\
\hat{B}_1 &= \frac{1}{\sqrt{2}} \left( \hat{\sigma}_2^x - \hat{\sigma}_2^z \right), \\
\hat{B}_2 &= \frac{1}{\sqrt{2}} \left( \hat{\sigma}_2^x + \hat{\sigma}_2^z \right),
\end{aligned} \quad (2.66)
$$

with Pauli matrices $\hat{\sigma}_i^\alpha$ acting on site $i$ of a two-site spin-1/2 system. With this, we define the specific CHSH-observable $\mathcal{B}^{\text{BP}}$,

$$
\begin{aligned}
\mathcal{B}^{\text{BP}} &= \frac{1}{\sqrt{2}} \left[ \langle \hat{\sigma}_1^x \hat{\sigma}_2^x \rangle - \langle \hat{\sigma}_1^x \hat{\sigma}_2^z \rangle + \langle \hat{\sigma}_1^x \hat{\sigma}_2^x \rangle + \langle \hat{\sigma}_1^x \hat{\sigma}_2^z \rangle \right. \\
&\qquad \left. + \langle \hat{\sigma}_1^z \hat{\sigma}_2^x \rangle - \langle \hat{\sigma}_1^z \hat{\sigma}_2^z \rangle - \langle \hat{\sigma}_1^z \hat{\sigma}_2^x \rangle - \langle \hat{\sigma}_1^z \hat{\sigma}_2^z \rangle \right] \quad (2.67) \\
&= \sqrt{2} \left[ \langle \hat{\sigma}_1^x \hat{\sigma}_2^x \rangle - \langle \hat{\sigma}_1^z \hat{\sigma}_2^z \rangle \right].
\end{aligned}
$$

We can choose the state vector of the Bell pair, $|\Psi^{\text{BP}}\rangle$, and calculate $\mathcal{B}^{\text{BP}}$ by plugging it in,

$$
\begin{aligned}
\mathcal{B}^{\text{BP}} &= \frac{1}{\sqrt{2}} \left[ \langle \uparrow \downarrow | + \langle \downarrow \uparrow | \right] \hat{\sigma}_1^x \hat{\sigma}_2^x \left[ |\uparrow \downarrow \rangle + |\downarrow \uparrow \rangle \right] \\
&\quad - \frac{1}{\sqrt{2}} \left[ \langle \uparrow \downarrow | + \langle \downarrow \uparrow | \right] \hat{\sigma}_1^z \hat{\sigma}_2^z \left[ |\uparrow \downarrow \rangle + |\downarrow \uparrow \rangle \right] \quad (2.68) \\
&= 2\sqrt{2},
\end{aligned}
$$

where we use

$$\hat{\sigma}^x|\uparrow\rangle = |\downarrow\rangle, \ \hat{\sigma}^x|\downarrow\rangle = |\uparrow\rangle,$$
$$\hat{\sigma}^z|\uparrow\rangle = |\uparrow\rangle, \ \hat{\sigma}^z|\downarrow\rangle = -|\downarrow\rangle, \tag{2.69}$$
$$\langle\uparrow|\uparrow\rangle = \langle\downarrow|\downarrow\rangle = 1, \ \langle\uparrow|\downarrow\rangle = \langle\downarrow|\uparrow\rangle = 0.$$

Here we find

$$|\mathcal{B}^{\mathrm{BP}}| = 2\sqrt{2} > 2. \tag{2.70}$$

Thus, the CHSH-inequality is violated. This is why we choose the upper index "BP" here, referring to an observable which violates the CHSH-inequality for the Bell pair. The violation shows that the Bell state is an entangled state and exhibits quantum correlations. It can hence not be described by the theory of locally-acting hidden variables [10, 11]. It has further been shown that $|\mathcal{B}| = 2\sqrt{2}$ is the upper limit of general CHSH-observables for quantum states, so that for the Bell state the CHSH-inequality is maximally violated [12].

Considering measurements on the Bell state, we find that the magnetizations in the $\hat{\sigma}^x$- and $\hat{\sigma}^z$-basis are both zero,

$$\langle\hat{\sigma}_i^x\rangle = \langle\hat{\sigma}_i^z\rangle = 0, \tag{2.71}$$

where the operators act on site $i \in \{1, 2\}$. The correlations between the two spins are given by

$$\langle\hat{\sigma}_1^x\hat{\sigma}_2^x\rangle = 1,$$
$$\langle\hat{\sigma}_1^z\hat{\sigma}_2^z\rangle = -1. \tag{2.72}$$

### 2.4.2  GHZ State

A generalization of a Bell state to larger spin systems was introduced by Greenberger, Horne and Zeilinger in 1989 and is accordingly called the Greenberger–Horne–Zeilinger (GHZ) state [13]. A Bell state can also be constructed with the state vector

$$|\tilde{\Psi}^{\mathrm{BP}}\rangle := \frac{1}{\sqrt{2}}\left(|\uparrow\uparrow\rangle + |\downarrow\downarrow\rangle\right), \tag{2.73}$$

as this is also a two-spin system in a superposition between two macroscopically different states. Thus, the two spins are strongly entangled, and the state maximally violates the CHSH-inequality, which can be checked straightforwardly by choosing the corresponding observables. A generalization to a system with $N > 2$ spins is then given by the state vector of the so-called GHZ state

$$|\Psi^{\text{GHZ}}\rangle := \frac{1}{\sqrt{2}} \left( |\uparrow\uparrow \ldots \uparrow\uparrow\rangle + |\downarrow\downarrow \ldots \downarrow\downarrow\rangle \right). \tag{2.74}$$

This is a superposition of the state with all $N$ spins being up and the state with all $N$ spins being down. The state is hence strongly entangled [13–15].

If we consider expectation values of measurements in the $z$-basis, one can straightforwardly see from the state vector that an even number of $\hat{\sigma}_i^z$-operators acting on different sites yields an expectation value of one. On the other hand, the action of an odd number of $\hat{\sigma}_i^z$-operators on different sites results in an expectation value of zero,

$$\langle \Psi^{\text{GHZ}}| \bigotimes_{i=1}^{K} \sigma_i^z |\Psi^{\text{GHZ}}\rangle = \frac{1}{2}\langle \uparrow\uparrow \ldots \uparrow\uparrow| \bigotimes_{i=1}^{K} \hat{\sigma}_i^z |\uparrow\uparrow \ldots \uparrow\uparrow\rangle$$

$$+ \frac{1}{2}\langle \downarrow\downarrow \ldots \downarrow\downarrow| \bigotimes_{i=1}^{K} \hat{\sigma}_i^z |\downarrow\downarrow \ldots \downarrow\downarrow\rangle \tag{2.75}$$

$$= \frac{1}{2} + \frac{1}{2}(-1)^K.$$

We can, without loss of generality, apply the $\hat{\sigma}_i^z$-operators to the spins on the first $K$ sites due to symmetry reasons.

In the $x$-basis, a $\hat{\sigma}_i^x$-operator flips the spin on site $i$ it is acting on. Thus, expectation values of operators in the $x$-basis can only be non-zero if a $\hat{\sigma}_i^x$-operator acts on all sites. In this case the expectation value is one,

$$\langle \Psi^{\text{GHZ}}| \bigotimes_{i=1}^{K} \hat{\sigma}_i^x |\Psi^{\text{GHZ}}\rangle = \begin{cases} 1 & \text{if } K = N, \\ 0 & \text{else.} \end{cases} \tag{2.76}$$

## 2.5  The Transverse-Field Ising Model

### 2.5.1  Ground-State Properties

The one-dimensional transverse-field Ising model (TFIM) is an integrable model for spin-1/2 particles. It is defined on a chain of $N$ sites with nearest-neighbor Ising interactions of strength $J$ in the $\hat{\sigma}^x$-basis in an external field of strength $h_{\text{t}}$ in the $\hat{\sigma}^z$-basis. The Hamiltonian reads

$$\hat{H}_{\text{TFIM}} = -J \sum_{j=1}^{N} \hat{\sigma}_j^x \hat{\sigma}_{(j+1)\text{mod}N}^x - h_{\text{t}} \sum_{j=1}^{N} \hat{\sigma}_j^z, \tag{2.77}$$

where $(j+1)\text{mod}N$ denotes the modulo $N$ calculation that accounts for periodic boundary conditions applied [16–21].

The model can be solved analytically via a Jordan–Wigner fermionization, where it is mapped onto non-interacting fermions which provide the exact spectrum and energy eigenstates [16, 21]. From this analytical solution it can be seen that the model describes a second-order phase transition, or quantum phase transition, at $h_t = h_{t,c} = \pm J$ between a paramagnetic ($|h_t| > J$) and a ferromagnetic phase ($|h_t| < J$). These quantum critical points are characterized by a gapless dispersion relation. The order parameter of the quantum phase transition is $\langle \hat{\sigma}^x \rangle$ [20, 22].

We sketch the Jordan–Wigner fermionization according to [16, 21, 23] and derive an expression for the dispersion relation to see the vanishing gap at the quantum critical points. Therefore, we first fix the energy scale by choosing $J = 1$, which can be done without loss of generality. Furthermore, we have to distinguish between chains with an even and an odd number of spins during the fermionization procedure. Here we only consider the case of an even number of sites, for which we show simulation results later. The calculations for an odd number of sites can be done analogously, but provide slightly different results, which however in the end lead to the same quantum critical points [16, 21, 23].

In the Jordan–Wigner fermionization framework, Pauli spin operators are mapped to fermionic Jordan–Wigner operators $\hat{a}_j$, $\hat{a}_j^\dagger$. These can be straightforwardly shown to satisfy the canonical commutation relations as stated in Eq. (2.24). The mapping can be done via

$$\hat{a}_j^\dagger := \exp\left[ i\pi \sum_{k=1}^{j-1} \hat{\sigma}_k^+ \hat{\sigma}_k^- \right] \hat{\sigma}_j^+, \tag{2.78}$$

$$\hat{a}_j = \hat{\sigma}_j^- \exp\left[ -i\pi \sum_{k=1}^{j-1} \hat{\sigma}_k^+ \hat{\sigma}_k^- \right], \tag{2.79}$$

with $\hat{\sigma}_j^\pm = 1/2(\hat{\sigma}_j^x \pm i\hat{\sigma}_j^y)$.

In terms of these fermionic operators, the Pauli spin operators can be expressed as

$$\hat{\sigma}_j^x = \exp\left[ i\pi \sum_{k=1}^{j-1} \hat{a}_k^\dagger \hat{a}_k \right] \left( \hat{a}_j^\dagger + \hat{a}_j \right), \tag{2.80}$$

$$\hat{\sigma}_j^z = \hat{\mathbb{1}} - 2\hat{a}_j^\dagger \hat{a}_j, \tag{2.81}$$

and the Hamiltonian becomes [16, 21, 23]

$$\hat{H}_{\text{TFIM}} = -\sum_{j=1}^{N} \left[ 2h_t \hat{a}_j^\dagger \hat{a}_j + \left( \hat{a}_j^\dagger - \hat{a}_j \right) \left( \hat{a}_{(j+1)\bmod N}^\dagger + \hat{a}_{(j+1)\bmod N} \right) \right]$$

$$+ h_t N + \left( \hat{a}_N^\dagger - \hat{a}_N \right) \left( \hat{a}_1^\dagger + \hat{a}_1 \right) \left[ \hat{\mathbb{1}} + \exp\left( i\pi \sum_{j=1}^{N} \hat{a}_j^\dagger \hat{a}_j \right) \right]. \tag{2.82}$$

Given this expression, a discrete Fourier transform can be applied to express the system in momentum space. For this we introduce the operators

$$\hat{b}_p := \frac{1}{\sqrt{N}} \sum_{j=1}^{N} \hat{a}_j \exp\left[ipj\right], \tag{2.83}$$

$$\hat{b}_p^\dagger = \frac{1}{\sqrt{N}} \sum_{j=1}^{N} \hat{a}_j^\dagger \exp\left[-ipj\right], \quad p \in \frac{\pi}{N}\mathbb{Z}, \tag{2.84}$$

where $p$ is the discrete Fourier mode. With a bit of algebra, the Hamiltonian can be expressed in momentum space as [16, 21, 23]

$$\begin{aligned}
\hat{H}_{\mathrm{TFIM}} =& -\sum_{p>0} \left[ (2h_t + 2\cos\left[p\right]) \left( \hat{b}_p^\dagger \hat{b}_p - \hat{b}_{-p} \hat{b}_{-p}^\dagger \right) \right. \\
& \left. +2i\sin\left[p\right] \hat{b}_{-p} \hat{b}_p - 2i\sin\left[p\right] \hat{b}_p^\dagger \hat{b}_{-p}^\dagger \right] \\
=& -\sum_{p>0} \left[ \hat{b}_p^\dagger \ \hat{b}_{-p} \right] \begin{bmatrix} 2\left(h_t + \cos\left[p\right]\right) & -2i\sin\left[p\right] \\ 2i\sin\left[p\right] & -2\left(h_t + \cos\left[q\right]\right) \end{bmatrix} \begin{bmatrix} \hat{b}_p \\ \hat{b}_{-p}^\dagger \end{bmatrix}.
\end{aligned} \tag{2.85}$$

The matrix form in the last line is used in order to diagonalize the Hamiltonian by a Bogoliubov transformation defined as

$$\hat{c}_p := u_p \hat{b}_p + v_p \hat{b}_{-p}^\dagger, \tag{2.86}$$

which yields new fermionic operators $\hat{c}_p$, $\hat{c}_p^\dagger$ sufficing Eq. (2.24). The coefficients $u_p$ and $v_p$ are given by

$$u_p = \begin{cases} 0 & \text{if } p \in 2\pi\left(\mathbb{Z} + \frac{1}{2}\right), \\ 1 & \text{if } p \in 2\pi\mathbb{Z}, \end{cases} \tag{2.87}$$

$$v_p = \begin{cases} -i & \text{if } p \in 2\pi\left(\mathbb{Z} + \frac{1}{2}\right), \\ 0 & \text{if } p \in 2\pi\mathbb{Z}. \end{cases} \tag{2.88}$$

In terms of these operators the Hamiltonian reads [16, 21, 23]

$$\hat{H}_{\mathrm{TFIM}} = \sum_p \omega_p \left( \hat{c}_p^\dagger \hat{c}_p + \frac{1}{2} \right), \tag{2.89}$$

with frequency

$$\omega_p = 2\sqrt{1 + h_t^2 - 2h_t \cos\left(p\right)}. \tag{2.90}$$

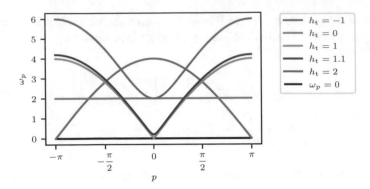

**Fig. 2.2** Plot of the dispersion relation $\omega_p$ (Eq. 2.90) as a function of momentum $p$ inside the Brillouin-zone for the TFIM with different transverse fields $h_t$. For comparison a horizontal line is drawn at $\omega_p = 0$ to visualize that the dispersion relation gets gapless for $h_t = \pm 1$. The gap closes in the middle of the Brillouin-zone for $h_t = 1$ and at the edges for $h_t = -1$, indicating quantum phase transitions at these two quantum critical points

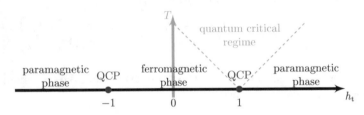

**Fig. 2.3** Phase diagram of the TFIM indicating the two quantum critical points (QCP) at $h_t = \pm 1$ at which quantum phase transitions between a ferromagnetic and a paramagnetic phase appear. These quantum phase transitions only exist at zero temperature, which is what we consider in this thesis. The phase diagram for non-zero temperatures is illustrated in gray for $h_t > 0$, as it is discussed in [20]

The Hamiltonian is now in a non-interacting diagonal form, so that $\omega_p$ describes the single-particle energy spectrum and thus the dispersion relation between frequency $\omega$ and momentum $p$ [16, 21, 23].

In Fig. 2.2 we plot this dispersion relation for different values of $h_t$ and it can be seen that it becomes gapless for $h_t = \pm 1$. For $h_t = 1$ the gap closes in the middle of the Brillouin-zone, while it closes at its edges for $h_t = -1$. Thus, at both transverse fields we find a quantum phase transition [20, 22, 23]. At these quantum critical points, the entanglement entropy reaches its maximum and scales exponentially with the system size [24]. Figure 2.3 shows the resulting phase diagram, where it needs to be noticed that the quantum phase transition only occurs at zero temperature, which is also the case we consider in the following [20, 22].

From the energy spectrum we can calculate the ground-state energy per particle $E_0$ for finite system sizes as [16, 21, 23]

$$E_0 = -\frac{1}{N} \sum_{p \in P_N} \sqrt{1 + h_t^2 - 2h_t \cos(p)}, \tag{2.91}$$

$$P_N = \frac{2\pi}{N} \left\{ -\frac{N-1}{2}, -\frac{N-3}{2}, \dots, \frac{N-1}{2} \right\}. \tag{2.92}$$

In the thermodynamic limit, $N \to \infty$, we find the ground-state energy per site $\varepsilon_0$ for infinitely long chains [16, 21, 23],

$$\varepsilon_0 = -\frac{1}{2\pi} \int_{-\pi}^{\pi} \sqrt{1 + h_t^2 - 2h_t \cos(p)} \, dp. \tag{2.93}$$

As the TFIM can be solved analytically and additionally shows quantum phase transitions, it is a toy model for benchmarking approximate simulation methods, since the simulation outcomes can be compared with the exact solution for arbitrary system sizes and in quantum critical regimes.

## 2.5.2 Dynamics After Sudden Quenches

In the TFIM dynamics after sudden quenches can be calculated analytically [17–19]. Here a sudden quench refers to the procedure of preparing the system in the ground state of the TFIM Hamiltonian with some initial transverse field $h_{t,i}$ and suddenly changing it to a final value $h_{t,f}$ at time $t = 0$. In the system where $h_t = h_{t,f}$, the prepared initial state is an out-of-equilibrium state that reveals non-equilibrium dynamics.

The dynamics in the $xx$-correlation function after such a quench in the TFIM,

$$C_d^{xx}(t, h_{t,i}, h_{t,f}) = \langle \hat{\sigma}_0^x \hat{\sigma}_d^x \rangle, \tag{2.94}$$

can be calculated analytically by evaluating the determinant of a matrix which we do not derive here, for more detailed information see [17, 19, 23]. Instead we state the final result,

$$
\begin{aligned}
C_d^{xx}(t, h_{t,i}, h_{t,f}) &= C_0(h_{t,i}, h_{t,f}) \exp\left[ -\frac{d}{\xi_1(h_{t,i}, h_{t,f})} \right] \\
&+ (h_{t,f}^2 - 1)^{1/4} \sqrt{4h_{t,f}} \int_{-\pi}^{\pi} \frac{dk}{\pi} \left[ \frac{n_{\mathrm{BF}}(k)}{1 - n_{\mathrm{BF}}(k)} \right]^{1/2} \frac{\sin\left[ 2\omega_{\mathrm{BF}}(k, h_{t,f}) t - kd \right]}{\omega_{\mathrm{BF}}(k, h_{t,f})} \\
&\times \exp\left( \int_0^{\pi} \frac{dk}{\pi} \ln |1 - 2n_{\mathrm{BF}}(k)| \right. \\
&\times \left. \left\{ d + \Theta\left[ d - 2v_{\mathrm{BF}}(k, h_{t,f}) t \right] \left[ 2v_{\mathrm{BF}}(k, h_{t,f}) t - d \right] \right\} \right),
\end{aligned}
\tag{2.95}
$$

with Heaviside step function $\Theta(x)$ and amplitude

$$C_0\left(h_{\mathrm{t,i}}, h_{\mathrm{t,f}}\right) = \left[\frac{\left(h_{\mathrm{t,i}} - h_{\mathrm{t,f}}\right) h_{\mathrm{t,f}} \sqrt{h_{\mathrm{t,i}}^2 - 1}}{\left(h_{\mathrm{t,i}} + h_{\mathrm{t,f}}\right)\left(h_{\mathrm{t,i}} h_{\mathrm{t,f}} - 1\right)}\right]^{1/2}, \qquad (2.96)$$

for $h_{\mathrm{t,i}} > h_{\mathrm{t,f}}$.

Furthermore, we introduce the inverse correlation length,

$$\xi_1^{-1}\left(h_{\mathrm{t,i}}, h_{\mathrm{t,f}}\right) = \Theta\left(h_{\mathrm{t,f}} - 1\right)\Theta\left(h_{\mathrm{t,i}} - 1\right)\ln\left|\min\left(h_{\mathrm{t,i}}, h_{\mathrm{t,1}}\right)\right| \\ - \frac{1}{2\pi}\int_{-\pi}^{\pi} dk \ln\left[1 - 2n_{\mathrm{BF}}(k)\right], \qquad (2.97)$$

$$h_{\mathrm{t,1}} = \frac{1 + h_{\mathrm{t,f}} h_{\mathrm{t,i}} + \sqrt{\left(h_{\mathrm{t,f}}^2 - 1\right)\left(h_{\mathrm{t,i}}^2 - 1\right)}}{h_{\mathrm{t,f}} + h_{\mathrm{t,i}}}, \qquad (2.98)$$

and the mode occupation numbers of the Bogoliubov fermions diagonalizing the TFIM Hamiltonian,

$$n_{\mathrm{BF}}\left(k, h_{\mathrm{t,i}}, h_{\mathrm{t,f}}\right) = \frac{1}{2} - 2\frac{h_{\mathrm{t,i}} h_{\mathrm{t,f}} + 1 - \left(h_{\mathrm{t,i}} + h_{\mathrm{t,f}}\right)\cos(k)}{\omega_{\mathrm{BF}}\left(k, h_{\mathrm{t,f}}\right)\omega_{\mathrm{BF}}\left(k, h_{\mathrm{t,i}}\right)}. \qquad (2.99)$$

The mode frequencies take the form as stated in Eq. (2.90),

$$\omega_{\mathrm{BF}}\left(k, h_{\mathrm{t}}\right) = 2\sqrt{h_{\mathrm{t}}^2 + 1 - 2h_{\mathrm{t}}\cos(k)}. \qquad (2.100)$$

From these we get the group velocity [17, 19, 23],

$$v_{\mathrm{BF}}\left(k, h_{\mathrm{t}}\right) = \frac{\mathrm{d}\omega_{\mathrm{BF}}\left(k, h_{\mathrm{t}}\right)}{\mathrm{d}k}. \qquad (2.101)$$

In the following we focus on the special case of quenches from a very large initial transverse field, $h_{\mathrm{t,i}} \to \infty$, into the vicinity of the quantum critical point within the paramagnetic regime, $h_{\mathrm{t,f}} \to 1$, $h_{\mathrm{t,f}} > 1$. This case has been studied in detail albeit the exact solution can be used for any initial and final transverse field as long as $h_{\mathrm{t,i}} > h_{\mathrm{t,f}}$ [17, 19, 23].

Since we cannot set the initial transverse field to infinity, we choose it large, $h_{\mathrm{t,i}} \geq 100$. We have checked that a fully $z$-polarized state, as would be the ground state for an infinitely large transverse field, is approximated accurately. For the ground state with large transverse field, the largest value in the $xx$-correlation is reached for nearest neighbors due to the correlation function showing an exponential decay as a function of the relative distance $d$. Numerical calculations give values around $C_{d=1}^{xx} \approx 5 \times 10^{-4}$ for the ground state at $h_{\mathrm{t}} = 100$, which is sufficiently close to zero within machine precision. Thus, the fully $z$-polarized state where all $xx$-correlations are zero is approximated with suitable accuracy [19].

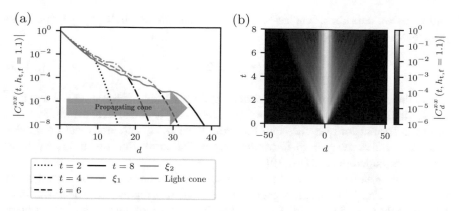

**Fig. 2.4** **a** Correlation function $C_d^{xx}$ (Eq. 2.95) as a function of relative distance $d$ at different times $t$ after a sudden quench to $h_{t,f} = 1.1$ in a chain with $N = 100$ sites and periodic boundary conditions. For small relative distances stationarity is reached at short times and an exponential decay defining a correlation length $\xi_1$ is found (blue regime). At larger relative distances oscillations appear whose envelope decays exponentially, defining a second correlation length $\xi_2$ (orange regime). At even larger relative distances a Gaussian fall-off is observed which propagates outwards with time (green regime). Outside the cone the correlations fall to zero (black regime). **b** Color plot of the correlation function $C_d^{xx}$ (Eq. 2.95) as a function of relative distance $d$ and time $t$ after a sudden quench to $h_{t,f} = 1.1$ in a chain with $N = 100$ sites and periodic boundary conditions. The propagating light cone is clearly visible, where inside the cone the correlations decay exponentially, while outside of it the correlations fall to zero. Figure adapted from [25]

Figure 2.4a shows the exact correlation function at different times after a sudden quench from $h_{t,i} = 1000$ to a final transverse field $h_{t,f} = 1.1$ as a function of the relative distance $d$ between the spins whose correlation is calculated. There it can be seen that for short relative distances, in the blue regime, the correlation function reaches stationarity already at very short times after the quench. At larger relative distances the correlations vanish, and one can see a Gaussian fall-off between the non-vanishing and the vanishing correlations, as indicated in green in Fig. 2.4a. The scale at which this Gaussian fall-off starts propagates outwards to larger relative distances. Hence, a light-cone-like behavior can be observed which is more clearly visible in Fig. 2.4b, showing the time evolution of the correlation function in a color plot as function of the relative distance $d$ [17, 19].

One can see that the propagation in time is roughly linear and the corresponding velocity is given by the propagation velocity of the elementary excitations in the system. This reaches its maximum for the maximum occupation number in the fermion distribution. Thus, the velocity of the propagating cone is given by the maximum group velocity, defined in Eq. (2.101). Its maximum is given by

$$v_{max}(h_t) = \max [v(k, h_t) | k \in [-\pi, \pi]]$$
$$= 2\min[h_t, 1], \tag{2.102}$$

and since we consider quenches within the paramagnetic phase, $h_t \geq 1$, the cone propagates with velocity [23]

$$v_{\text{cone}} \simeq 2. \tag{2.103}$$

Inside the cone, one can see a stationary exponential fall-off for short relative distances, indicated in blue in Fig. 2.4a. This stationary behavior at short relative distances is special for quenches into the vicinity of the quantum critical point. For quenches further away from it, oscillations in the correlation function have been found to be superimposed [17, 19].

At larger relative distances, time-dependent oscillations can be found, indicated in orange in Fig. 2.4a. The propagation of the oscillating regime is much slower than the one of the Gaussian fall-off. Considering the envelope of the oscillating regime in the correlation function after a sudden quench close to the quantum critical point, it also shows an exponential fall-off with relative distance $d$. Thus, we can extract two correlation lengths. The first one, $\xi_1(t, h_{t,f})$, is defined by the exponential decay at small relative distances. The second one, $\xi_2(t, h_{t,f})$, we can extract from the envelope of the oscillating regime at larger relative distances [17, 19, 23].

In the following we focus on the first correlation length, which we can extract from the correlation function by fitting an exponential function to the short-distance regime, $d \leq 3$,

$$C_d^{xx}(t, h_{t,f}) \propto \exp\left[ -\frac{d}{\xi_1(t, h_{t,f})} \right]. \tag{2.104}$$

As for quenches into the vicinity of the quantum critical point stationarity is reached already after short times, the correlation length rather depends on the final transverse field $h_{t,f}$ than on time. The stationary correlation length is given by the diagonal elements of the initial-state density matrix, which is set up from the fermion expectation numbers. These depend only on the initial and final transverse field, as they remain stationary after the sudden quench [19]. Thus, the matrix takes the form of a so-called generalized Gibbs ensemble (GGE) and in the case of a very large initial transverse field the stationary correlation length is found to follow the GGE behavior described by the universal function [17–19, 26, 27]

$$\xi_{\text{GGE}}(h_{t,f}) = \left[ \ln(2h_{t,f}) \right]^{-1}. \tag{2.105}$$

A GGE is a generalization of a canonical ensemble which describes classical systems in thermal equilibrium. A system with constant particle number in a fixed volume inside a heat bath at thermal equilibrium is represented by a Gibbs ensemble in (classical) statistical physics. In this representation the statistical entropy of a canonical state is maximized under the constraint of energy conservation and the state can be described by a Boltzmann distribution. This property can be translated to quantum mechanics via replacing the statistical entropy by the von-Neumann

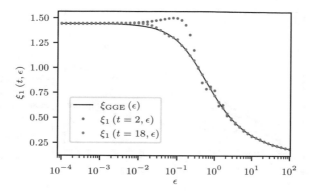

**Fig. 2.5** Correlation length $\xi_1$ [Eq. (2.104)] at times $t = 2$ and $t = 18$ after a sudden quench to different distances $\epsilon$ [Eq. (2.106)] from the quantum critical point in a chain with $N = 100$ sites and periodic boundary conditions. The expected behavior at stationarity described by a generalized Gibbs ensemble (GGE) is shown for comparison according to Eq. (2.105). While stationarity is reached already at short times for small and large distances from the quantum critical point, longer times are necessary to reach stationarity at intermediate $\epsilon$. There the GGE behavior appears only in the long-time limit. Figure adapted from [25]

entropy of the density matrix, introduced in Eq. (2.60). However, it is maximized under more constraints of conserved quantities than only the energy, causing the expression "generalized" Gibbs ensemble [26, 27].

In Fig. 2.5 the GGE behavior of the correlation length according to Eq. (2.105) is plotted as a function of the distance,

$$
\epsilon = \frac{h_{\text{t,f}} - h_{\text{t,c}}}{h_{\text{t,c}}}
$$
$$
= h_{\text{t,f}} - 1, \tag{2.106}
$$

of the final transverse field from the quantum critical point at $h_{\text{t,c}} = 1$. It is compared with correlation lengths extracted from the exact correlation function at different times $t$ after a sudden quench in a chain with $N = 100$ sites. It can be seen that for quenches close to the quantum critical point stationarity is reached already after very short times. At larger distances from the quantum critical point long times are necessary until the correlation length gets stationary and follows the GGE behavior. Thus, in this regime the GGE behavior is an asymptotic long-time limit [17–19, 23].

However, it is interesting to see that already after short times the correlation function at short relative distances shows the near-critical behavior expected at late times after the quench. Thus, these results can even be used to probe thermal equilibrium critical properties already at short times [19]. This is an important and useful property, since the quantum critical regime is hard to capture with existing simulation methods due to the entanglement growing linearly as a function of time after the sudden quench.

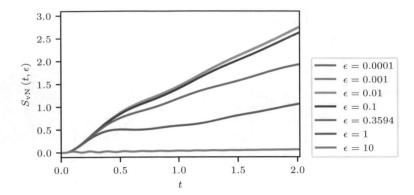

**Fig. 2.6** Half-chain von-Neumann entanglement entropy $S_{vN}(t, \epsilon)$ [Eq. (2.60)] as a function of time after sudden quenches to different distances $\epsilon$ [Eq. (2.106)] from the quantum critical point $h_{t,c} = 1$ in a spin chain with $N = 40$ sites. The entanglement entropy grows linearly with time for quenches close to the quantum critical point, where the first three lines lie on top of each other, while it grows slower further away from the quantum phase transition. The entanglement entropy is calculated using time-dependent density-matrix renormalization group (tDMRG). Figure adapted from [28]

Figure 2.6 shows the von-Neumann entanglement entropy, Eq. (2.60), calculated for a chain with $N = 40$ sites split in two equally-sized subsystems, as a function of time after sudden quenches to different distances $\epsilon$ from the quantum critical point. The von-Neumann entanglement entropy is calculated using time-dependent density-matrix renormalization group (tDMRG), a simulation method working still well at these times and system sizes, see Sect. 2.6 [29].

One finds that the entanglement entropy grows linearly with time close to the quantum critical point. Comparing different chain lengths, one finds that the entanglement entropy saturates in this regime at values increasing extensively with the system size, showing volume-law behavior. Further away from the quantum phase transition the entanglement entropy grows much slower. Thus, large entanglement is reached already at short times after quenches into the quantum critical regime. This is a limit for many existing simulation methods, such as tDMRG, so that in such cases only short-time dynamics are accessible [29].

### 2.5.3   Adding a Longitudinal Field

Integrability is lost when applying an additional longitudinal field of strength $h_l$ in the $\hat{\sigma}^x$-basis to the TFIM [30], a model that we abbreviate with LTFIM. The Hamiltonian then reads

$$\hat{H}_{\text{LTFIM}} = -J \sum_{j=1}^{N} \hat{\sigma}_j^x \hat{\sigma}_{(j+1)\text{mod}N}^x - h_{\text{t}} \sum_{j=1}^{N} \hat{\sigma}_j^z - h_{\text{l}} \sum_{j=1}^{N} \hat{\sigma}_j^x. \tag{2.107}$$

We can again fix the energy scale by setting $J = 1$.

We still find the system in a paramagnetic phase for large $h_{\text{t}}$ and small $h_{\text{l}}$ and in a ferromagnetic phase for small $h_{\text{t}}$, but between the phases there is no quantum phase transition anymore. The ferromagnetic state is not degenerate here due to the longitudinal field and thus no symmetry is broken when going to the paramagnetic regime [30].

If we again consider sudden quenches from a large transverse field and no longitudinal field, ($h_{\text{t,i}} = 100$, $h_{\text{l,i}} = 0$), to different final values, ($h_{\text{t,f}}, h_{\text{l,f}} \neq 0$), as we did in the TFIM, we expect oscillations in the correlation length. These we expect to result from Rabi oscillations between the spin states caused by the interaction of the longitudinal and the transverse field.

As the model is not integrable anymore, we can only access an exact solution for very small system sizes, where the Hamiltonian can be diagonalized exactly. For larger system sizes we need to deal with approximate numerical simulations instead.

## 2.6  Exact Diagonalization and tDMRG

In contrast to the TFIM, most spin systems, like the LTFIM, are not integrable and hence no closed-form solution can be found. Therefore, to get an impression of the system behavior, such models require numerical approximations which are well understood and known to represent the exact solution accurately in the considered regimes.

For small system sizes, up to $N \sim 10 - 20$ sites, a complete exact diagonalization of the Hamiltonian operator can still be applied numerically. Thus, we can get exact solutions for the whole state spectrum and the dynamics [31]. A complete diagonalization of the Hamiltonian operator provides the eigenstates and eigenenergies, which corresponds to solving the stationary Schrödinger equation,

$$\hat{H}|\Psi\rangle = E|\Psi\rangle. \tag{2.108}$$

The ground state is then given by the eigenstate with the smallest eigenenergy and can be expressed in a suitable basis.

To calculate the time evolution after a sudden quench, we can evolve each eigenstate via

$$|\Psi(t)\rangle = e^{-i\hat{H}t}|\Psi(0)\rangle, \tag{2.109}$$

which can be calculated given the diagonalized Hamiltonian [31]. Computing the exact diagonalization for the system is very expensive but enables the possibility to calculate all desired properties.

The diagonalization can be computed more efficiently if symmetries reducing the Hilbert space dimension are taken into account. Additionally, several numerical approaches exist to iteratively diagonalize the Hamiltonian in an efficient way, such as the Lanczos and the Jacobi–Davidson algorithm [31–33], or the Krylov methods for dynamics [31]. These provide approximations of the exact result with good accuracy. With these approximate methods, only ground states or lowest-lying excited states can be calculated, which is sufficient in most cases. By applying appropriate operators, dynamics can still be calculated from the ground state only [31–33].

As the Hilbert space grows exponentially with system size, this diagonalization ansatz is still limited to small spin systems even though such approximate methods exist. To deal with larger system sizes, we need different approximation schemes. We can make use of the fact that in many cases not all states in the Hilbert space are physical states, which we call states the system can be found in with significant probabilities. Neglecting spin states with sufficiently small probabilities provides a reasonable approximation of the system. This reduces the effective Hilbert space dimension and shifts the problem to finding an efficient way to extract those physical states. Such an ansatz cannot approximate arbitrary states efficiently, but only those for which the number of physical states scales polynomially with the system size. However, for these states methods based on this ansatz turn out to be very helpful since they capture the non-generic properties of the quantum system [29, 34–37].

Here we consider specifically the (t)DMRG, which is a simulation method for one-dimensional quantum many-body systems whose precision is controllable via truncation. The principle ansatz of DMRG to find ground states is an iterative procedure, starting with considering only two sites at the borders of the system and increasing the chain length block-wise. Thus, in each iteration step, a part of the chain is considered consisting of two equally-sized blocks. One site is then added to each block at the inner end, shrinking the distance between the two, as illustrated in Fig. 2.7. This iteratively increases the considered chain length. The Hamiltonian operator resulting from the combination of both blocks is then diagonalized numerically to find the eigenstates of the system. A variational ansatz can be used for this to make the calculations cheaper [29, 34, 35, 38, 39].

From these eigenstates, the density matrix can be calculated, and it can be truncated by applying a singular-value decomposition. This factorizes a matrix into two unitary matrices and a diagonal matrix containing the singular values. Taking only the $D$ largest singular values into account and setting the others to zero truncates the range of the matrix. The decomposition can then be reversed again to get the reduced density matrix. This reduces the range of the density matrix and the effective dimension of the Hilbert space under consideration cannot exceed the so-called bond dimension $D$. Due to the splitting algorithm in the DMRG process, this bond dimension can be controlled dynamically. Moreover, information about the error made in the truncation can be extracted, telling how to adapt the bond dimension [29, 34–36].

After this truncation, the basis of the considered system is changed by redefining the two blocks including the additional sites and two more sites of the chain are added between the blocks. The iteration process is illustrated in Fig. 2.7. If the combined size of the two blocks reaches the actual chain length, the iteration can still be proceeded by sweeping from left to right and back. The size of one block is then increased while decreasing the size of the other block, where all sets need to be kept track of [29, 34, 35].

The variational search for the eigenstates can be expressed in the matrix product state (MPS) formulation, which is an efficient representation of low-energy states in one-dimensional quantum systems [40]. The coefficients $c_v$ in the basis expansion of the state vector can be understood as elements of a rank-$N$ tensor for a spin system with $N$ sites. Using a singular-value decomposition, that tensor can be successively split into local tensors $\tilde{A}_i^{(v_i)}$ of rank one acting on particle $i$ in state $|v_i\rangle$. One then obtains the Schmidt decomposition of the full tensor, where the local tensors are connected via diagonal matrices containing the Schmidt coefficients. These diagonal matrices can be pulled into the local tensors and this combination can be denoted as rank-one tensor $A_i^{(v_i)}$ to yield the MPS formulation [29, 36, 41, 42],

$$
\begin{aligned}
|\Psi\rangle &= \sum_{\{\mathbf{v}\}} c_{\mathbf{v}} |\mathbf{v}\rangle \\
&= \sum_{\{\mathbf{v}\}} A_1^{(v_1)} A_2^{(v_2)} \ldots A_N^{(v_N)} |\mathbf{v}\rangle.
\end{aligned}
\tag{2.110}
$$

For translational-invariant systems this expression gets simplified since $A_i^{(v_i)} = A^{(v_i)}$ for all $i \in \{1, \ldots, N\}$. The connecting matrices containing the Schmidt coefficients quantify the entanglement between the two particles whose connection got cut. If not all Schmidt coefficients are non-zero, which is the case for low-energy states with small entanglement, the MPS representation can be truncated by neglecting the (close to) zero Schmidt coefficients. This provides an ansatz to efficiently approximate quantum many-body systems with good accuracy [36].

It can directly be seen that this MPS form underlies the block form used in DMRG and an equivalence between MPS and DMRG is given in the thermodynamic limit [38]. The explicit form of the matrices can be determined via a variational ansatz and provides the eigenstates of a block in the DMRG procedure [29].

The MPS expression in Eq. (2.110) is true for systems with open boundary conditions, while a connection between the first and last spin sites needs to be added when considering periodic boundary conditions. This is computationally more expensive, so that open boundary conditions can be simulated more efficiently in the MPS representation. It has been shown that a bond dimension $D$ in open boundary simulations corresponds to a bond dimension $D^2$ in periodic boundary simulations to reach comparable accuracy [29, 35].

To simulate dynamics, a Trotter-Suzuki decomposition can be used, where the time-evolution operator is approximated via a decomposition into operators $\hat{H}_j$ acting on individual links between sites $j$ and $j + 1$ in an $N$-site spin chain [29, 41, 43],

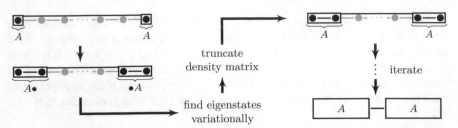

**Fig. 2.7** Schematic visualization of the DMRG approach. Arrows denote the workflow. The dots represent the spins on the chain, where the ones not taking part in the DMRG at this step are gray. Boxes denote the blocks which are increased by one site in each iteration and which are directly connected, neglecting the gray sites in between. One full iteration step consists of adding one site to each block $A$, giving $A\cdot$ and $\cdot A$, finding the eigenstates via a variational ansatz, calculating and truncating the density matrix and changing the basis, where the new blocks are defined as the old blocks plus the added site. This is repeated until the actual system size is reached and the whole chain is contained in the two blocks. At this point more iterations can be applied to increase the accuracy via shifting sites from one block to the other, sweeping through the chain, where all possible block-shifts need to be considered [29, 34, 35]

$$
\exp\left[-it\hat{H}\right] \approx \prod_{j=1}^{N-1} \exp\left[-it\hat{H}_j\right]
$$

$$
= \exp\left[-it\frac{\hat{H}_1}{2}\right] \exp\left[-it\frac{\hat{H}_2}{2}\right] \ldots \exp\left[-it\hat{H}_{N-1}\right] \ldots \quad (2.111)
$$

$$
\times \ldots \exp\left[-it\frac{\hat{H}_2}{2}\right] \exp\left[-it\frac{\hat{H}_1}{2}\right].
$$

This decomposition is often referred to as trotterization. Using the MPS representation of the states, Eq. (2.110), an operator acting only on sites $j$ and $j+1$ can be applied to the wave function exactly in the DMRG step where the two blocks are separated by the sites $j$ and $j+1$. Thus, each of the individual operators can be applied exactly in the corresponding DMRG step. The action of the trotterized time-evolution operator can then be calculated via a sweep from site one to site $N$ and back [43]. This is the extension of DMRG to tDMRG, enabling the simulation of dynamics, for example after sudden quenches.

The DMRG approximation works best in regimes where the eigenvalues decay exponentially, as this makes a truncation possible without losing too much information [29, 34, 35]. However, if the eigenvalues decay only slowly, a truncation is harder, since many eigenstates play an important role in the system and cannot be neglected. This behavior is found if the individual sites are strongly entangled. Thus, the entanglement entropy is a quantity giving an estimate how large the bond dimension needs to be chosen. In the worst case, the system is strongly entangled, and the necessary bond dimension grows exponentially with the system size, so that the DMRG becomes inefficient [29, 34–36].

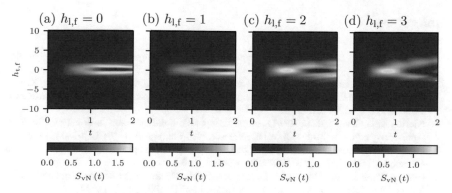

**Fig. 2.8** Von-Neumann entanglement entropy $S_{vN}(t)$ [Eq. (2.112)] as a function of time and final transverse field after sudden quenches from initial fields $h_{t,i} = 100$, $h_{l,i} = 0$ to longitudinal fields $h_{l,f} = 0$ [(a)], $h_{l,f} = 1$ [(b)], $h_{l,f} = 2$ [(c)] and $h_{l,f} = 3$ [(d)] in a system with $N = 42$ sites. The entanglement entropy is calculated via tDMRG where for regimes of strong entanglement larger bond dimensions need to be chosen to get accurate results, which is the case in all panels for regimes of small transverse fields. Figure adapted from [28]

The same limitation applies to the dynamics in tDMRG. If the entanglement grows linearly in time, the tDMRG is limited to short times and an exponentially growing bond dimension is necessary to capture the dynamics at later times [29, 43].

Figure 2.8 shows the von-Neumann entanglement entropy,

$$S_{vN} = - \operatorname{Tr}\left[\hat{\rho}_A \log \hat{\rho}_A\right], \tag{2.112}$$

with half-chain reduced density matrix $\hat{\rho}_A$, as a function of time after sudden quenches in the LTFIM. This entanglement entropy can be calculated within the tDMRG procedure. The dynamics are shown after sudden quenches at time $t = 0$ from an effectively fully $z$-polarized state, $h_{t,i} = 100$, $h_{l,i} = 0$, to final longitudinal fields $h_{l,f} = 0$ [Fig. 2.8a], $h_{l,f} = 1$ [Fig. 2.8b], $h_{l,f} = 2$ [Fig. 2.8c] and $h_{l,f} = 3$ [Fig. 2.8d]. The final transverse field $h_{t,f}$ is varied along the $y$-axis.

For quenches in the TFIM [Fig. 2.8a], volume-law entanglement can be found at the quantum critical points, as already discussed in the context of Fig. 2.6. If a longitudinal field is present, the entanglement entropy also grows in time for quenches to small transverse fields, while it remains small at larger transverse fields. Thus, to simulate sudden quenches to small transverse fields a large bond dimension is necessary in the tDMRG calculations.

Having introduced the basics of quantum mechanics we do not use hats to indicate operators anymore in the following. It should now be clear from the context which quantities are operators.

# References

1. Troyer M (2012) Lecture notes on computational quantum physics. http://edu.itp.phys.ethz. ch/fs12/cqp/
2. Bartelmann Matthias, Lüst Dieter, Wipf Andreas, Rebhan Anton, Feuerbacher Björn, Krüger Timm (2015) Die Entstehung der Quantenphysik. Springer, Berlin, Heidelberg. https://doi.org/ 10.1007/978-3-642-54618-1_21
3. Nielsen MA, Chuang IL (2010) Quantum computation and quantum information. In: 10th, Anniversary edn. Cambridge University Press. https://doi.org/10.1017/CBO9780511976667
4. Parkinson JB, Farnell DJJ (2010) An introduction to quantum spin systems. Springer, Berlin Heidelberg. https://doi.org/10.1007/978-3-642-13290-2
5. Gerlach Walther, Stern Otto (1922) Der experimentelle Nachweis der Richtungsquantelung im Magnetfeld. Z Phys 9(1):349–352 Dec. https://doi.org/10.1007/BF01326983
6. Einstein A, Podolsky B, Rosen N (1935) Can quantum-mechanical description of physical reality be considered complete? Phys Rev 47:777–780 May. https://link.aps.org/doi/10.1103/ PhysRev.47.777
7. John Stewart Bell (1964) On the einstein podolsky Rrsen paradox. Physics 1(3):195–200. https://cds.cern.ch/record/111654
8. Bell John S (1966) On the problem of hidden variables in quantum mechanics. Rev Mod Phys 38:447–452 Jul. https://link.aps.org/doi/10.1103/RevModPhys.38.447
9. Bell JS, Alain A (2004) Speakable and unspeakable in quantum mechanics: collected papers on quantum philosophy, 2 edn. Cambridge University Press. https://doi.org/10.1017/ CBO9780511815676
10. Clauser John F, Horne Michael A, Shimony Abner, Holt Richard A (1969) Proposed experiment to test local hidden-variable theories. Phys Rev Lett 23:880–884 Oct. https://link.aps.org/doi/ 10.1103/PhysRevLett.23.880
11. Clauser John F, Horne Michael A (1974) Experimental consequences of objective local theories. Phys Rev D 10:526–535 Jul. https://link.aps.org/doi/10.1103/PhysRevD.10.526
12. Cirel'son BS (1980) Quantum generalizations of Bell's inequality. Lett Math Phys 4(2):93–100 Mar. https://doi.org/10.1007/BF00417500
13. Greenberger DM, Horne MA, Zeilinger A (1989) Going beyond Bell's theorem, pp 69–72. Springer, Netherlands, Dordrecht. https://doi.org/10.1007/978-94-017-0849-4_10
14. Dür W, Vidal G, Cirac JI (2000) Three qubits can be entangled in two inequivalent ways. Phys Rev A 62:062314 Nov. https://link.aps.org/doi/10.1103/PhysRevA.62.062314
15. Gisin N, Bechmann-Pasquinucci H (1998) Bell inequality, Bell states and maximally entangled states for $n$ qubits. Phys. Lett. A 246(1):1–6. http://www.sciencedirect.com/science/article/pii/ S0375960198005167
16. Pfeuty P (1970) The one-dimensional Ising model with a transverse field. Ann Phys 57:79–90. https://doi.org/10.1016/0003-4916(70)90270-8
17. Calabrese P, Essler FHL, Fagotti M (2012) Quantum quench in the transverse field Ising chain: I. time evolution of order parameter correlators. J Stat Mech Theory Exp 2012(07):P07016. https://doi.org/10.1088%2F1742-5468%2F2012%2F07%2Fp07016
18. Calabrese P, Essler FHL, Fagotti M (2012) Quantum quenches in the transverse field Ising chain: II. stationary state properties. J Stat Mech Theory Exp 2012(07):P07022. https://doi. org/10.1088%2F1742-5468%2F2012%2F07%2Fp07022
19. Karl Markus, Cakir Halil, Halimeh Jad C, Oberthaler Markus K, Kastner Michael, Gasenzer Thomas (2017) Universal equilibrium scaling functions at short times after a quench. Phys Rev E 96:022110 Aug. https://link.aps.org/doi/10.1103/PhysRevE.96.022110
20. Sachdev S (2011) Quantum phase transitions, 2 edn. Cambridge University Press. https://doi. org/10.1017/CBO9780511973765
21. Lieb Elliott, Schultz Theodore, Mattis Daniel (1961) Two soluble models of an antiferro-magnetic chain. Ann Phys 16(3):407–466. http://www.sciencedirect.com/science/article/pii/ 0003491661901154

22. Sachdev S, Young AP (1997) Low temperature relaxational dynamics of the Ising chain in a transverse field. Phys Rev Lett 78:2220–2223. https://link.aps.org/doi/10.1103/PhysRevLett.78.2220
23. Cakir H (2015) Dynamics of the transverse field Ising chain after a sudden quench. Master's thesis, Ruprecht-Karls-Universität Heidelberg
24. Iglói F, Lin Y-C (2008) Finite-size scaling of the entanglement entropy of the quantum Ising chain with homogeneous, periodically modulated and random couplings. J Stat Mech Theory Exp 2008(06):P06004. https://doi.org/10.1088/1742-5468/2008/06/p06004
25. Czischek S, Gärttner M, Oberthaler M, Kastner M, Gasenzer T (2018) Quenches near criticality of the quantum Ising chain–power and limitations of the discrete truncated Wigner approximation. Quant Sci Technol 4(1):014006. http://stacks.iop.org/2058-9565/4/i=1/a=014006
26. Jaynes ET (1957) Information theory and statistical mechanics. Phys Rev 106:620–630 May. https://link.aps.org/doi/10.1103/PhysRev.106.620
27. Jaynes ET (1957) Information theory and statistical mechanics. II. Phys Rev 108:171–190 Oct. https://link.aps.org/doi/10.1103/PhysRev.108.171
28. Czischek Stefanie, Gärttner Martin, Gasenzer Thomas (2018) Quenches near Ising quantum criticality as a challenge for artificial neural networks. Phys Rev B 98.024311 Jul. https://doi.org/10.1103/PhysRevB.98.024311
29. Schollwöck Ulrich (2011) The density-matrix renormalization group in the age of matrix product states. Ann Phys 326(1):96–192. http://www.sciencedirect.com/science/article/pii/S0003491610001752
30. Ovchinnikov AA, Dmitriev DV, Krivnov V Ya, Cheranovskii VO (2003) Antiferromagnetic Ising chain in a mixed transverse and longitudinal magnetic field. Phys Rev B 68(21):214406 Dec. https://link.aps.org/doi/10.1103/PhysRevB.68.214406
31. Noack RM, Manmana SR (2005) Diagonalization-and numerical renormalization-group-based methods for interacting quantum systems. AIP Conf Proc 789(1):93–163. https://aip.scitation.org/doi/abs/10.1063/1.2080349
32. Laflorencie N, Poilblanc D (2004) Simulations of pure and doped low-dimensional spin-1/2 gapped systems, pp 227–252. Springer, Berlin, Heidelberg. https://doi.org/10.1007/BFb0119595
33. Weiße A, Fehske H (2008) Exact diagonalization techniques, pp 529–544. Springer, Berlin, Heidelberg. https://doi.org/10.1007/978-3-540-74686-7_18
34. White Steven R (1992) Density matrix formulation for quantum renormalization groups. Phys Rev Lett 69:2863–2866 Nov. https://link.aps.org/doi/10.1103/PhysRevLett.69.2863
35. White Steven R (1993) Density-matrix algorithms for quantum renormalization groups. Phys Rev B 48:10345–10356 Oct. https://link.aps.org/doi/10.1103/PhysRevB.48.10345
36. Bridgeman JC, Chubb CT (2017) Hand-waving and interpretive dance: an introductory course on tensor networks. J Phys Math Theor 50(22):223001. https://doi.org/10.1088%2F1751-8121%2Faa6dc3
37. Vidal Guifré (2004) Efficient simulation of one-dimensional quantum many-body systems. Phys Rev Lett 93:040502 Jul. https://link.aps.org/doi/10.1103/PhysRevLett.93.040502
38. Östlund Stellan, Rommer Stefan (1995) Thermodynamic limit of density matrix renormalization. Phys Rev Lett 75:3537–3540 Nov. https://link.aps.org/doi/10.1103/PhysRevLett.75.3537
39. Baxter RJ (1968) Dimers on a rectangular lattice. J Math Phys 9(4):650–654. https://doi.org/10.1063/1.1664623
40. Orús Román (2014) A practical introduction to tensor networks: Matrix product states and projected entangled pair states. Ann Phys 349:117–158. https://www.sciencedirect.com/science/article/pii/S0003491614001596
41. Stoudenmire EM, White SR (2010) Minimally entangled typical thermal state algorithms. New J Phys 12(5):055026. https://doi.org/10.1088%2F1367-2630%2F12%2F5%2F055026
42. Dukelsky J, Martín-Delgado MA, Nishino T, Sierra G (1998) Equivalence of the variational matrix product method and the density matrix renormalization group applied to spin chains. EPL 43(4):457–462 Aug. https://doi.org/10.1209%2Fepl%2Fi1998-00381-x

43. White Steven R, Feiguin Adrian E (2004) Real-time evolution using the density matrix renormalization group. Phys Rev Lett 93:076401 Aug. https://link.aps.org/doi/10.1103/PhysRevLett.93.076401

# Chapter 3
# Artificial Neural Networks

During the last years much effort was expended to combine (quantum) physics with machine learning methods. This has led to many useful and interesting results. One of those was an ansatz to parametrize quantum spin-1/2 systems with a generative artificial neural network, specifically the restricted Boltzmann machine. We further analyze this ansatz within this thesis.

In this chapter we therefore introduce the basic rudiments of machine learning techniques based on artificial neural networks. This provides the foundation to introduce the parametrization ansatz of spin-1/2 systems. Albeit not all the concepts discussed here are employed further within this thesis, we give a detailed overview. The concepts of feed-forward neural networks are discussed in Sect. 3.1 based on [1–3], which provide an introduction to artificial neural networks. Section 3.2 discusses the restricted Boltzmann machine in detail and is based on [1–4], while in Sect. 3.3.1 we introduce supervised learning according to [1–3, 5]. The basics of unsupervised learning are discussed in Sect. 3.3.2 based on [1–4, 6] and we consider reinforcement learning in Sect. 3.3.3 according to [2, 7].

Besides the introduction of the basic methods, we give a review of applications of machine learning methods in (quantum) physics in Sect. 3.4, which is based on two detailed reviews given in [3, 7]. We end the chapter with discussing the neuromorphic hardware present in the BrainScaleS group at Heidelberg University, which emulates an artificial neural network on an analog hardware with hardwired spiking neurons. We are interested in combining this hardware with the quantum state parametrization based on artificial neural networks.

S. Czischek, *Neural-Network Simulation of Strongly Correlated Quantum Systems*, Springer Theses, https://doi.org/10.1007/978-3-030-52715-0_3

## 3.1   Discriminative Models: Feed-Forward Neural Networks

Nowadays, machine learning has become a famous approach in many regimes of technology and has found a wide range of applications, where it has shown significant improvements. To name a few examples, machine learning led to impressive progress in the fields of autonomous driving, text or speech recognition, and playing computer games. A famous application is also Google's AlphaGo, a machine which bet the European champion Fan Hui [8], as well as world champion Lee Sedol in the game "Go". All these applications are based on the task to recognize patterns in huge amounts of data [7].

While many approaches and methods of machine learning exist, which perform differently depending on the tasks considered, methods based on artificial neural networks (ANN) show remarkable results. This is especially true since people started to compose deep networks showing more internal structure. An illustrating example for the power of these models is image recognition, where the network can decide whether a cat (or any other object) can be found in a given picture. For this example, the ANN takes the pixels of a given image as input and propagates the information through a network. This is done in a way such that all information is mapped to a single output, telling whether there is a cat in the picture or not. Given some labeled data, the way of propagating the information can be modified such that a desired output is created, which is referred to as training the network. Once trained, the ANN can be applied to new input data and still provides the right classification [7].

The original idea of the ANN ansatz is to find a way to mimic the information processing found by biologists in brains of humans and animals via a mathematical description. However, on the way to the ANN models used nowadays, the biological plausibility got lost in most cases, as it turned out that the practicality of the mathematical ansatz implies restrictions which do not appear in real brains. Nevertheless, these restrictions allow ANN approaches to become practical models for performing machine learning with a remarkably efficient behavior.

A famous ANN ansatz is the feed-forward neural network (fNN) model, which we introduce as an example in this section. The fNN is the most general example for the class of discriminative models, which take some input and create an output that is observed. These models appear useful in applications such as regression, where values of target variables are predicted given some input variables, or classification, where an input vector is assigned to a discrete class. We introduce fNN networks and then discuss modifications of the network structure providing different discriminative ANN setups. Another class of ANN are generative models, such as the restricted Boltzmann machine, where a full model distribution is represented by the network. Thus, also new data points in the input space can be created. We discuss these models in more detail in Sect. 3.2.

The setup of an fNN is shown in Fig. 3.1a, where the network consists of many neurons, denoted by the circles, which can take continuous real values. The neurons can be grouped into different layers. Here neurons within one layer are placed below

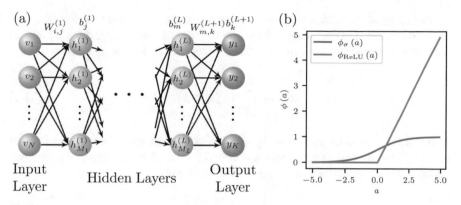

**Fig. 3.1** **a** Schematic representation of an fNN with an input layer consisting of $N$ input neurons $\mathbf{v}$, $L$ hidden layers consisting of $M_l$ hidden neurons $\mathbf{h}^{(l)}$ each and an output layer with $K$ output neurons $\mathbf{y}$. Information is processed from left to right, as indicated by the arrows, where each neuron gets as input $a_j^{(l)}$ the value of each neuron $h_i^{(l-1)}$ of the previous layer multiplied by some weight $W_{i,j}^{(l)}$ plus some bias $b_j^{(l)}$. **b** Two common choices for activation functions are the sigmoid function $\phi_\sigma(a) = 1/(1 + \exp[-a])$ and the rectified linear unit (ReLU) function $\phi_{\mathrm{ReLU}}(a) = 0$, if $a < 0$, and $\phi_{\mathrm{ReLU}}(a) = a$, else. Both are plotted as a function of neuron input $a$ to show that they reach values larger than zero for $a \geq 0$, which leads to an activation of the neuron. The sigmoid function is already non-zero for negative $a$ close to zero, but there it takes small values, so that it is a smooth version of the Heaviside step function

each other, such that each vertical set depicts one layer. The leftmost set is the input layer, whose neurons represent the given input data. The rightmost set is the output layer, containing the prediction of the network. In the example of the cat classification in an image, this output layer would consist of a single neuron telling whether there is a cat in the picture or not. The layers in between are referred to as hidden layers, whose neurons have no clear interpretation. Nevertheless, these layers are important for increasing the performance of the algorithm, where it has been found that more hidden layers increase the reachable accuracy of the network in predicting an output. One commonly also refers to the neurons as input, hidden or output variables and in this thesis we use these expressions synonymously.

The network structure is then defined via connections between the neurons of different layers. In an fNN, each neuron $v_i$ of the input layer with $N$ input neurons, $\mathbf{v} = (v_1, \ldots, v_N)$, is connected to each neuron $h_j^{(1)}$ of the first hidden layer with $M_1$ hidden neurons, $\mathbf{h}^{(1)} = (h_1^{(1)}, \ldots, h_{M_1}^{(1)})$, via weights $W_{i,j}^{(1)}$. Every neuron $h_j^{(1)}$ in the first hidden layer has an additional bias $b_j^{(1)}$. Analogously, each neuron $h_j^{(1)}$ of the first hidden layer is connected to every neuron $h_k^{(2)}$ of the second hidden layer with $M_2$ neurons, $\mathbf{h}^{(2)} = (h_1^{(2)}, \ldots, h_{M_2}^{(2)})$, via weights $W_{j,k}^{(2)}$, and so on for the following hidden layers. The biases are analogously present for all neurons in all hidden layers. The last hidden layer is connected to the output layer with $K$ output neurons, $\mathbf{y} = (y_1, \ldots, y_K)$.

The values of the hidden variables are calculated via non-linear functions depending on the neurons in the previous layer and the weights connecting them, as well as the biases. Considering a neuron in the first hidden layer, $h_j^{(1)}$, it gets the input

$$a_j^{(1)} = \sum_{i=1}^{N} W_{i,j}^{(1)} v_i + b_j^{(1)}, \qquad (3.1)$$

which is the sum over all incoming weights multiplied with the connected input variables plus the bias of the hidden neuron. The value of the neuron is then derived via applying an activation function $\phi(a)$ onto its input,

$$h_j^{(1)} = \phi\left[a_j^{(1)}\right]. \qquad (3.2)$$

Common choices for activation functions are the sigmoid function,

$$\phi_\sigma(a) = \frac{1}{1 + \exp[-a]}, \qquad (3.3)$$

or the rectified linear unit (ReLU) function,

$$\phi_{\text{ReLU}}(a) = \begin{cases} 0, & \text{if } a < 0, \\ a, & \text{else.} \end{cases} \qquad (3.4)$$

Both are plotted in Fig. 3.1b, where it can be seen that they are zero for $a < 0$, or small for $a$ negative but close to zero, and non-zero otherwise. Thus, the hidden neuron gets activated if its input crosses the threshold of zero.

The hidden neurons in the following layer are activated by the activation functions depending on the connections to the previous layer, and so on for further layers. The values of the neurons in the output layer are finally given by the total network function,

$$y_k = \phi\left[b_k^{(L+1)} + \sum_{j=1}^{M_L} W_{k,j}^{(L+1)} \phi\left(\dots \phi\left[b_m^{(1)} + \sum_{i=1}^{N} W_{m,i}^{(1)} v_i\right]\dots\right)\right], \qquad (3.5)$$

where we consider a network with $L$ hidden layers. The values of the hidden variables are plugged in iteratively as the activation function $\phi(a)$ of the input $a$, as stated in Eq. (3.1), until the dependence on the input layer is reached. Thus, information is processed through the network from the input to the output neurons via the hidden layers, causing the name feed-forward neural network. This is indicated by the arrows in Fig. 3.1a.

In total, the output neurons are highly non-linear functions depending on the input neurons with the weights and biases being adjustable parameters. These free parameters are adapted such that the network provides the desired output, where the procedure of finding the right parameters is called the training or learning of the network. Several methods to perform this training exist and can be chosen depending on the considered task. We discuss the training procedures in detail in Sect. 3.3.

It has been mathematically proven that an fNN can approximate any function arbitrarily well due to its high non-linearity, given enough hidden neurons. This can become inefficient and computationally expensive if too many hidden layers are needed, but in principle a suitable network can be trained for any input data set.

The choice of weights to get a certain output is not unique, as many symmetries can be found in the weight space. For example, the hidden neurons can be interchanged and the weights can be chosen such that the network still provides the same outcome. This is also true for changing the sign of the weights. Thus, many possible choices of weights provide the same network output and it is hard to interpret the weights found during training.

We have introduced the fNN model in a general way with an arbitrary number of hidden layers and also arbitrary numbers of neurons per layer. These numbers, as well as the activation functions chosen for the individual layers, are ingredients which define the network structure. They build a set of hyper-parameters which can be adapted suitably to find a well-performing network but cannot be derived generally. Due to the simple correspondence of the network structure to the underlying mathematical formula for the network function, the setup can straightforwardly be generalized further to more complex structures.

While the fNN ansatz is a commonly used one, there exist many other discriminative ANN models with different properties. To name a few, convolutional neural networks use kernels which are scanned over the input data and the weights of these kernels are learned in the training procedure. This way, features which are learned at one position in the data can also be recognized in other parts of it, so that translational invariance is directly included in the network structure. This is for example useful in face recognition tasks.

Another example are recurrent neural networks, which have an internal memory. Thus, each neuron can also depend on its state when given some previous input data. This can for example be helpful when analyzing or training words or sentences for text generation, where each word or letter also depends on the previous data. These networks are furthermore useful when considering robot controls, where each action depends on the previous one.

## 3.2   Generative Models: The Restricted Boltzmann Machine

### 3.2.1   Setup and Properties

Having introduced the discriminative ANN models, we now consider generative ANN models. These encode probability distributions underlying the neurons in the network. From such networks also data in the input space can be created, showing the

Hidden
Layer

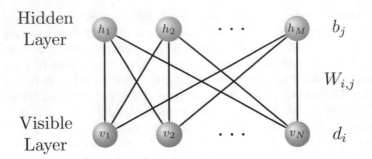

Visible
Layer

**Fig. 3.2** Schematic visualization of a two-layer restricted Boltzmann machine with $N$ binary visible variables $v_i$ and $M$ binary hidden variables $h_j$. Each visible variable is connected to each hidden variable via a weight $W_{i,j}$ and each neuron has an additional bias, $d_i$ for the visible and $b_j$ for the hidden neurons. The weights and biases provide the network energy which is considered to describe a Boltzmann distribution

same structure as the data the network is trained on. An example for an application of such a generative ANN would be completing fragmentary input data. The network can for example be trained on a set of hand-written digits, where it learns the probability of each pixel in the image to be black or white when a certain digit is in the picture. The network can then be used to fill up an image where a part of a written digit is missing, since new data can be created from the learned probabilities for the pixels in the input space.

As a general example of generative ANN models, we consider the restricted Boltzmann machine (RBM). This consists of two layers of neurons, the $N$ visible neurons, $\mathbf{v} = (v_1, \ldots, v_N)$, and the $M$ hidden neurons, $\mathbf{h} = (h_1, \ldots, h_M)$. We again synonymously use the expressions visible and hidden variables for the corresponding neurons within this thesis. A scheme of the RBM setup is shown in Fig. 3.2. Here the visible layer corresponds to the input layer in the fNN models discussed in Sect. 3.1 and contains the observed data. While every neuron $v_i$ in the visible layer is connected to each neuron $h_j$ in the hidden layer via some weight $W_{i,j}$, no connections within the layers exist. This restriction to inter-layer connections is reflected by the term "restricted" in the name of the model. It leads to various simplifications in dealing with the network compared to a general Boltzmann machine with all-to-all interactions. Each variable furthermore has a bias, analogously to the fNN model, where the visible variables have biases $d_i$ and the hidden variables have biases $b_j$.

The visible and hidden neurons in the RBM are usually both chosen binary and we here make the choice that they can take the variables $v_i, h_j \in \{\pm 1\}$. There are also works where the variables are chosen to be zero or one, but both choices can be transferred into each other. The RBM can also be expressed with variables taking general discrete values, which turns out to be useful for some applications. This can be reached by combining multiple binary neurons to one neuron with more possible states [9].

Considering the binary variables and the setup of the RBM, one can find an analogy to physical spin-1/2 systems, in particular to the Ising model in a field. Since the particles in a spin-1/2 system are binary, they can be identified with the neurons in the RBM. The spin-interactions in the Ising model can be mapped to the weights in the RBM. Thereby the biases of the neurons correspond to an external magnetic field acting on each spin. Thus, training the RBM can be interpreted as learning the values of interactions and magnetic fields in an Ising model.

In thermal equilibrium the spin system converges to a state which is described by a Boltzmann distribution. Hence, the probability distribution over the state space of the spin-1/2 system converges to

$$P\left(\mathbf{s}\right) = \frac{1}{Z} \exp\left[ -\frac{E\left(\mathbf{s}\right)}{k_{\mathrm{B}} T} \right], \tag{3.6}$$

$$Z = \sum_{\{\mathbf{s'}\}} \exp\left[ -\frac{E\left(\mathbf{s'}\right)}{k_{\mathrm{B}} T} \right], \tag{3.7}$$

with spin configurations $\mathbf{s}$, system energy $E(\mathbf{s})$, Boltzmann factor $k_{\mathrm{B}}$ and temperature $T$. $Z$ denotes the normalization factor with the sum running over all possible spin configurations. This explains why the considered ANN model is called (restricted) Boltzmann machine, since we expect the configurations of the neurons to follow a Boltzmann distribution in thermal equilibrium.

Using this analogy to the Ising model, the energy of a network configuration can be defined as

$$E\left(\mathbf{v}, \mathbf{h}; \mathcal{W}\right) = -\sum_{i=1}^{N} d_i v_i - \sum_{j=1}^{M} b_j h_j - \sum_{i=1}^{N}\sum_{j=1}^{M} v_i W_{i,j} h_j, \tag{3.8}$$

with the set of all weights and biases $\mathcal{W} = (\mathbf{d}, \mathbf{b}, W)$. Since the RBM is defined independent of the temperature, we can put $k_{\mathrm{B}} T = 1$, so that the probability distribution over the network configurations can be expressed as [7]

$$P\left(\mathbf{v}, \mathbf{h}; \mathcal{W}\right) := \frac{\tilde{P}\left(\mathbf{v}, \mathbf{h}; \mathcal{W}\right)}{Z\left(\mathcal{W}\right)} \tag{3.9}$$

$$= \frac{\exp\left[-E\left(\mathbf{v}, \mathbf{h}; \mathcal{W}\right)\right]}{Z\left(\mathcal{W}\right)},$$

$$Z\left(\mathcal{W}\right) = \sum_{\{\mathbf{v}\},\{\mathbf{h}\}} \exp\left[-E\left(\mathbf{v}, \mathbf{h}; \mathcal{W}\right)\right]. \tag{3.10}$$

Here we distinguish between the unnormalized distribution $\tilde{P}(\mathbf{v}, \mathbf{h}; \mathcal{W})$ and the distribution $P(\mathbf{v}, \mathbf{h}; \mathcal{W})$ normalized by the factor $Z(\mathcal{W})$. This is already the output we expect from a generative model, providing a probability distribution underlying all neurons. However, the most common application of an RBM is to model probability distributions. Hence, the learning procedure consists of adapting the weights such that a probability distribution underlying some input data is reproduced. This distribution is in most cases only defined for the visible variables, as these are the given input parameters of the network and hence correspond to samples following the target distribution. Thus, we also need to know $P(\mathbf{v}; \mathcal{W})$ in order to minimize its difference from the desired distribution during the network training.

The quantity $P(\mathbf{v}; \mathcal{W})$ is given by the marginalization over the hidden variables,

$$P(\mathbf{v}; \mathcal{W}) = \frac{1}{Z(\mathcal{W})} \sum_{\{\mathbf{h}\}} \tilde{P}(\mathbf{v}, \mathbf{h}; \mathcal{W}). \qquad (3.11)$$

Here the sum runs over all $2^M$ possible configurations of hidden variables. Analogously, the distribution underlying the hidden variables is given by a marginalization over the visible variables,

$$P(\mathbf{h}; \mathcal{W}) = \frac{1}{Z(\mathcal{W})} \sum_{\{\mathbf{v}\}} \tilde{P}(\mathbf{v}, \mathbf{h}; \mathcal{W}). \qquad (3.12)$$

Calculating this distribution again includes a sum over exponentially many terms, the $2^N$ configurations of the visible variables. The so-called block Gibbs sampling scheme, which we introduce further in Sect. 3.2.2, can be used to iteratively create samples of the visible and hidden variables from the underlying distributions. These can approximate the sums in Eqs. (3.11)–(3.12), so that the probability distributions can be calculated efficiently in an approximate manner.

An unsupervised learning scheme as introduced in Sect. 3.3.2 can be used to train the weights and biases in the RBM such that $P(\mathbf{v}; \mathcal{W})$ represents a probability distribution underlying the given input data accurately. Therefore, the target distribution is not needed to be known explicitly. By drawing samples from $P(\mathbf{v}; \mathcal{W})$ via a Monte Carlo approach, new data can be created in the input space, which represents the structure of the given input data. Analogously to the fNN, it has been shown that the RBM can represent any probability distribution given enough hidden neurons [10].

By stacking multiple RBM models on top of each other, a so-called deep Boltzmann machine with multiple hidden layers can be created. Thereby the hidden layer of the lower RBM acts as a visible layer for the RBM on top of it. This increases the number of variational parameters in the network and thus improves its representational power.

## 3.2.2  Gibbs Sampling

To sample both, the visible and the hidden neurons in an RBM, a common choice is the Gibbs sampling approach. This is an ansatz that can be applied efficiently to networks with no intra-layer interactions. It makes use of the fact that neurons of one layer depend only on neurons of a different layer and thus each layer can be fully updated at once to create a new sample.

The Gibbs sampling scheme is a Markov chain Monte Carlo method, meaning that a chain of samples is created with each new sample depending only on the previous one. Here the visible and hidden variables are updated alternatingly. Thus, the iterative sampling procedure of a two-layer network with one visible and one hidden layer takes the form

$$\cdots \to (\mathbf{v}_t, \mathbf{h}_t) \to (\mathbf{v}_{t+1}, \mathbf{h}_t) \to (\mathbf{v}_{t+1}, \mathbf{h}_{t+1}) \to \ldots, \tag{3.13}$$

where $t$ is the index counting the sampling steps.

Considering the RBM setup, the unnormalized joint probability of a configuration of neurons is given by $\tilde{P}(\mathbf{v}, \mathbf{h}; \mathcal{W})$, as defined in Eq. (3.9). From this we can derive transition probabilities for the visible variables to change from $\mathbf{v}$ to $\tilde{\mathbf{v}}$ while keeping the hidden variables fixed,

$$
\begin{aligned}
T\left[(\mathbf{v}, \mathbf{h}) \to (\tilde{\mathbf{v}}, \mathbf{h})\right] &\equiv T(\mathbf{v} \to \tilde{\mathbf{v}}) \\
&:= \frac{\tilde{P}(\tilde{\mathbf{v}}, \mathbf{h}; \mathcal{W})}{\sum_{\{\tilde{\mathbf{v}}\}} \tilde{P}(\tilde{\mathbf{v}}, \mathbf{h}; \mathcal{W})} \\
&= P(\tilde{\mathbf{v}} | \mathbf{h}; \mathcal{W}),
\end{aligned}
\tag{3.14}
$$

where we introduce the conditional probability $P(\mathbf{v}|\mathbf{h}; \mathcal{W})$. From this expression we can derive the acceptance probability $A(\mathbf{v} \to \tilde{\mathbf{v}})$. In the creation of the sample chain, some configuration $\mathbf{v}$ is given and a new configuration $\tilde{\mathbf{v}}$ is proposed. This new configuration is then either accepted with probability $A(\mathbf{v} \to \tilde{\mathbf{v}})$, so that $\tilde{\mathbf{v}}$ is added to the sample set, or rejected with probability $1 - A(\mathbf{v} \to \tilde{\mathbf{v}})$ so that $\mathbf{v}$ is added to the sample set. The acceptance probability reads

$$
\begin{aligned}
A(\mathbf{v} \to \tilde{\mathbf{v}}) &= \min\left[1, \frac{\tilde{P}(\tilde{\mathbf{v}}, \mathbf{h}; \mathcal{W})}{\tilde{P}(\mathbf{v}, \mathbf{h}; \mathcal{W})} \frac{T(\tilde{\mathbf{v}} \to \mathbf{v})}{T(\mathbf{v} \to \tilde{\mathbf{v}})}\right] \\
&= 1,
\end{aligned}
\tag{3.15}
$$

where we use Eq. (3.14) to arrive at the last line. This tells us that each proposed configuration is accepted.

The new configuration can be sampled from the transition probability $T(\mathbf{v} \to \tilde{\mathbf{v}})$, as this provides samples following the desired distribution. Thus, for the sampling procedure we need the conditional probabilities,

$$P\left(\mathbf{v}\,|\mathbf{h};\,\mathcal{W}\right) = \frac{\tilde{P}\left(\mathbf{v},\mathbf{h};\,\mathcal{W}\right)}{\sum_{\{\mathbf{v}'\}}\tilde{P}\left(\mathbf{v}',\mathbf{h};\,\mathcal{W}\right)}$$

$$= \prod_{i=1}^{N} \frac{\exp\left[\sum_{j=1}^{M} v_i W_{i,j} h_j + d_i v_i\right]}{\sum_{v_i'=\pm1}\exp\left[\sum_{j=1}^{M}\tilde{v}_i W_{i,j} h_j + \tilde{v}_i d_i\right]} \qquad (3.16)$$

$$= \prod_{i=1}^{N} P\left(v_i\,|\mathbf{h};\,\mathcal{W}\right),$$

showing that the individual visible variables are conditionally independent. This provides the conditional probabilities for each visible neuron,

$$P\left(v_i = 1\,|\mathbf{h};\,\mathcal{W}\right) = \frac{1}{1 + \exp\left[-2\sum_{j=1}^{M} W_{i,j} h_j - 2d_i\right]}, \qquad (3.17)$$

$$P\left(v_i = -1\,|\mathbf{h};\,\mathcal{W}\right) = \frac{1}{1 + \exp\left[2\sum_{j=1}^{M} W_{i,j} h_j + 2d_i\right]}. \qquad (3.18)$$

Using the analogous steps, we can also write down the conditional probabilities for the hidden neurons,

$$P\left(h_j = 1\,|\mathbf{v};\,\mathcal{W}\right) = \frac{1}{1 + \exp\left[-2\sum_{i=1}^{N} v_i W_{i,j} - 2b_j\right]}, \qquad (3.19)$$

$$P\left(h_j = -1\,|\mathbf{v};\,\mathcal{W}\right) = \frac{1}{1 + \exp\left[2\sum_{i=1}^{N} v_i W_{i,j} + 2b_j\right]}. \qquad (3.20)$$

We can then use a standard importance sampling scheme to sample from the conditional distributions, where we iteratively update the visible and the hidden variables. This yields the following procedure for the Gibbs sampling scheme,

1. start with random configurations for the hidden variables,
2. generate $N$ uniformly distributed random variables $r_i \in [0, 1)$,
3. if $r_i \leq P(v_i = 1|\mathbf{h};\,\mathcal{W})$ set $\tilde{v}_i = 1$, else set $\tilde{v}_i = -1$ for all $i \in \{1, \ldots, N\}$,
4. generate $M$ uniformly distributed random numbers $r_j \in [0, 1)$,
5. if $r_j \leq P(h_j = 1|\tilde{\mathbf{v}};\,\mathcal{W})$ set $\tilde{h}_j = 1$, else set $\tilde{h}_j = -1$ for $j \in \{1, \ldots, M\}$,
6. add the created samples to the sample sets for visible and hidden neurons and repeat from 2 until a desired size of the sample sets is reached.

When dealing with deep networks, we can make use of the fact that each layer is only connected to the directly neighboring layers, so that layer one does not directly depend on layer three and so on. Thus, we can always create samples of the neurons in all odd layers at once and given those we can create samples of the neurons in all even layers at once, yielding two steps in the iteration scheme. The Gibbs sampling is hence an efficient sampling scheme when considering neural networks in the RBM structure.

Given this efficient sampling procedure, an RBM can be used to perform stochastic inference, as it can be trained to represent a desired probability distribution and samples can be drawn efficiently from it.

## 3.3  Training Neural Networks

### 3.3.1  Supervised Learning

Training neural networks is the procedure of adjusting the network weights and biases such that a desired output of the network is reached for fNN models, or such that a desired probability distribution is represented by an RBM [4]. Depending on the problem under consideration, different methods for training the network exist. These can be grouped as supervised learning, unsupervised learning and reinforcement learning. In the examples of classification or regression, where labeled data is given and the task of the network is to label unknown data, a supervised learning scheme is used [7].

Given some labeled data set, we can train an fNN to predict labels for unseen data according to some structure in the input. In the case of regression, the data has continuous real labels, so that training the network basically corresponds to fitting a function to data. The activation function of the output layer is then an identity function [7].

In the case of classification, the data labels take discrete values, so that each data point is allocated to one of the possible classes. The activation function of the output layer is then a so-called softmax function, which is a generalization of the sigmoid function to multiple dimensions,

$$\phi\left(a_k\right) = \frac{\exp\left[a_k\right]}{\sum_{j=1}^{K} \exp\left[a_j\right]}. \tag{3.21}$$

The sum in the denominator runs over all variables in the layer considered. Here these are the $K$ output variables, with each output neuron representing one of the possible classes. This way, the output layer provides a probability for the input data to belong to each class.

To train the network, some training data $\mathbf{v}$ needs to be given, which already carries labels $\mathbf{y}_l$. The network parameters can be adapted such that the difference between a label predicted by the network for a given input data $v_i$, $y_i = f_{\mathcal{W}}(v_i)$, and the true label, $y_{l,i} = f(v_i)$, is minimized for all data in the training set. The predicted label depends on all weights and biases in the network, which we summarize in the set $\mathcal{W}$. Thus, we can define a so-called cost function

$$C(\mathcal{W}) = \frac{1}{2} \left\langle \| f_{\mathcal{W}}(v) - f(v) \|^2 \right\rangle$$

$$\approx \frac{1}{2S} \sum_{i=1}^{S} \| f_{\mathcal{W}}(v_i) - f(v_i) \|^2 , \tag{3.22}$$

where the sum in the second line runs over all $S$ data points in the training data set, approximating the exact expectation value. This cost function is the sum-of-squares function and by minimizing it we can find network weights and biases such that the labels in the training data are best represented.

The most common way to train the fNN via minimizing the cost function is to use the gradient descent method. The idea of this ansatz is to follow the negative gradient of the cost function, which equals sliding down the hill in parameter space. This converges to the weights representing a minimum of the cost function.

If we consider one weight (or bias) $\mathcal{W}_\lambda$ and change it by some small amount $\delta \mathcal{W}_\lambda$,

$$\mathcal{W}_\lambda \to \mathcal{W}_\lambda + \delta \mathcal{W}_\lambda, \tag{3.23}$$

the cost function changes accordingly as

$$C(\mathcal{W}) \to C(\mathcal{W} + \delta \mathcal{W}_\lambda) = C(\mathcal{W}) + \delta_\lambda C(\mathcal{W}), \tag{3.24}$$

$$\delta_\lambda C(\mathcal{W}) := \delta \mathcal{W}_\lambda \frac{\partial}{\partial \mathcal{W}_\lambda} C(\mathcal{W}). \tag{3.25}$$

Stationarity is reached for $\partial C(\mathcal{W})/\partial \mathcal{W}_\lambda = 0$, which corresponds to an optimum or a saddle point.

From this it follows that the weights can be adapted according to

$$\mathcal{W}_\lambda^{(\tau+1)} = \mathcal{W}_\lambda^{(\tau)} - \varepsilon \frac{\partial}{\partial \mathcal{W}_\lambda} C\left(\mathcal{W}^{(\tau)}\right), \tag{3.26}$$

where $\tau$ counts the iteration steps when updating the weights. Moreover, $\varepsilon$ is a step size of the update, called the learning rate, which needs to be chosen appropriately. This formula denotes the gradient descent scheme. After each update of the weights, the cost function needs to be re-evaluated.

Calculating the gradient of the cost function is a non-trivial task for the highly non-linear network function $f_{\mathcal{W}}(v)$. An efficient way to calculate this gradient in the fNN is the so-called back-propagation procedure. Based on the chain rule, this scheme provides a single backward propagation through the network giving all derivatives [7].

To derive a way to evaluate the derivative of the cost function with respect to the network weights, we consider an arbitrary weight $W_{i,j}^{(l)}$ connecting neuron $i$ from layer $l-1$ with neuron $j$ from layer $l$. We take into account that the cost function only depends on the weight via the input $a_j^{(l)}$ which is fed into neuron $h_j^{(l)}$, so that we can write

$$\frac{\partial C\,(\mathcal{W})}{\partial W_{i,j}^{(l)}} = \frac{\partial C\,(\mathcal{W})}{\partial a_j^{(l)}} \frac{\partial a_j^{(l)}}{\partial W_{i,j}^{(l)}}. \tag{3.27}$$

Here we consider a connecting weight, but an analogous expression can be written down for the biases. As the input $a_j^{(l)}$ is a weighted sum of the neurons $h_k^{(l-1)}$ in the previous hidden layer,

$$a_j^{(l)} = \sum_{k=1}^{M_{l-1}} W_{k,j}^{(l)} h_k^{(l-1)} + b_j^{(l)}, \tag{3.28}$$

we get

$$\frac{\partial a_j^{(l)}}{\partial W_{i,j}^{(l)}} = h_i^{(l-1)}. \tag{3.29}$$

By applying the chain rule, one gets an iterative expression for the derivative of the cost function with respect to the neuron input,

$$\frac{\partial C\,(\mathcal{W})}{\partial a_j^{(l)}} = \sum_{k=1}^{M_{l+1}} \frac{\partial C\,(\mathcal{W})}{\partial a_k^{(l+1)}} W_{j,k}^{(l+1)} \frac{\partial \phi\left(a_j^{(l)}\right)}{\partial a_j^{(l)}}. \tag{3.30}$$

Here $\phi(a)$ is the activation function which needs to be chosen continuously so that the derivative can be calculated. Hence, the derivative of the cost function can be calculated given the expressions for the derivatives with respect to the inputs of all neurons in the following layer. Considering the output layer, $l = L + 1$, the label predicted by the network is given as the activation function applied to the output layer. Thus, for this case the derivative of the cost function is given by the absolute deviation between the predicted and the true label.

This way, starting from the explicit expression for the output layer, the error can be propagated backwards through the network to calculate the update rules for all weights and biases, where one pass yields all derivatives. Therefore, this procedure is called the error back-propagation scheme. In total, the weight update in the supervised learning thus consists of two steps. First the derivatives of the cost function are calculated via back-propagation and then all weights are updated according to the gradient descent procedure.

While the gradient descent is a reasonable ansatz, it still performs poorly in many cases. The main problem is that the parameter space is highly complex, so that many local minima and saddle-points exist in which the gradient descent can get stuck. We are interested in finding the global minimum of the cost function, but due to symmetries in the weight space, as discussed in Sect. 3.1, also the global minimum is highly degenerate. Thus, it is helpful to run the gradient descent multiple times starting from different initial points and compare the minima found.

Another ansatz to make the gradient descent more efficient is to use the so-called stochastic gradient descent. There not all data points in the training set are used at once to calculate the cost function, but only a few are considered at a time. These sets of few data points are called mini-batches and the gradient descent procedure according to Eq. (3.26) can be applied on them. To calculate the gradients, the back-propagation scheme is applied for each mini-batch individually and in the end the results are summed up. Using this ansatz, it is still possible to get stuck in local optima or saddle points, but it is more likely to escape from those. The gradients in each update step point in different directions since the corresponding cost function can vary depending on the mini-batches, adding stochasticity to the gradient descent algorithm.

In the gradient descent scheme in Eq. (3.26) we introduce the learning rate $\varepsilon$, which is another hyper-parameter next to the number of layers or neurons per layer, so it needs to be chosen appropriately. While a too small learning rate leads to very slow convergence, with a too large learning rate the update procedure can jump across the minimum in parameter space, leading to lower accuracy and divergences. Choosing the right learning rate is often a hard task and it is in many cases helpful to choose an adaptive learning rate, for example an exponentially decaying one. For this example, the convergence is fast in the beginning, and at the end the minimum is found with good accuracy.

### 3.3.2   Unsupervised Learning

An unsupervised learning scheme can be applied if no labeled data is given, as in the case of the RBM discussed in Sect. 3.2. This scheme can for example be used to find clusters in data or to model a probability distribution from which samples can be drawn. These are then statistically similar to the input data. The ansatz finds applications in de-noising, filling in of missing data or discrimination [7].

Considering the RBM, the goal of training the neural network is to represent a given probability distribution in the visible layer. Analogously to the case of supervised learning, to get an update for the network weights and biases we need to define a cost function which we can minimize. In the example of modeling a probability distribution, we thus need to find a difference between two distributions, which should be minimal for a well-trained network. Such a difference is given by the Kullback–Leibler divergence, which is defined as

$$D_{\mathrm{KL}}\left[P\left(\mathbf{v}\right) \| Q\left(\mathbf{v}\right)\right] := \sum_{\{\mathbf{v}\}} P\left(\mathbf{v}\right) \log\left[\frac{P\left(\mathbf{v}\right)}{Q\left(\mathbf{v}\right)}\right]. \tag{3.31}$$

$P(\mathbf{v})$ is the target distribution over a data set $\{\mathbf{v}\}$ and $Q(\mathbf{v})$ is the represented distribution. This Kullback–Leibler divergence tends to zero if the represented distribution approaches the target distribution and is otherwise positive, so that it can be considered as a distance measure between the distributions. It is important to notice that the Kullback–Leibler divergence is not symmetric, the target and the represented distribution cannot be exchanged.

Considering the RBM with the underlying distribution of the visible variables as stated in Eq. (3.11),

$$
P(\mathbf{v}; \mathcal{W}) = \frac{1}{Z(\mathcal{W})} \sum_{\{\mathbf{h}\}} \exp\left[ \sum_{i=1}^{N}\sum_{j=1}^{M} v_i W_{i,j} h_j + \sum_{i=1}^{N} d_i v_i + \sum_{j=1}^{M} h_j b_j \right],
$$
$$
Z(\mathcal{W}) = \sum_{\{\mathbf{h}\}}\sum_{\{\mathbf{v}\}} \exp\left[ \sum_{i=1}^{N}\sum_{j=1}^{M} v_i W_{i,j} h_j + \sum_{i=1}^{N} d_i v_i + \sum_{j=1}^{M} h_j b_j \right],
$$

(3.32)

and the target distribution $P_t(\mathbf{v})$ underlying the input data, we can write the cost function as the Kullback–Leibler divergence of the two,

$$
\begin{aligned}
C(\mathcal{W}) &= D_{\mathrm{KL}}\left[ P_t(\mathbf{v}) \| P(\mathbf{v}; \mathcal{W}) \right] \\
&= \sum_{\{\mathbf{v}\}} P_t(\mathbf{v}) \log\left[ \frac{P_t(\mathbf{v})}{P(\mathbf{v}; \mathcal{W})} \right].
\end{aligned}
$$

(3.33)

While minimizing this cost function provides a reasonable weight update, considering a slightly different cost function leads to the more efficient contrastive divergence (CD) learning, which is commonly used. Instead of directly minimizing the difference between the target and the represented probability distribution, the tendency of running away from the target distribution when drawing a single sample via Gibbs sampling can be minimized. This provides a reasonable ansatz, since in the Gibbs sampling scheme the probability of the sampled states cannot decrease during the chain creation, see Sect. 3.2.2. Thus, we define the cost function as the difference between two Kullback–Leibler divergences, the one of the represented distribution at equilibrium $P(\mathbf{v}; \mathcal{W})$ and the target distribution $P_t(\mathbf{v})$, and the one of the represented distribution at equilibrium $P(\mathbf{v}; \mathcal{W})$ and after a single sampling step, $P^{(1)}(\mathbf{v}; \mathcal{W})$. The cost function then reads

$$
\begin{aligned}
C_{\mathrm{CD}}(\mathcal{W}) &= D_{\mathrm{KL}}\left[ P_t(\mathbf{v}) \| P(\mathbf{v}; \mathcal{W}) \right] - D_{\mathrm{KL}}\left[ P^{(1)}(\mathbf{v}; \mathcal{W}) \| P(\mathbf{v}; \mathcal{W}) \right] \\
&= \sum_{\{\mathbf{v}\}} P_t(\mathbf{v}) \log\left[ \frac{P_t(\mathbf{v})}{P(\mathbf{v}; \mathcal{W})} \right] - \sum_{\{\mathbf{v}\}} P^{(1)}(\mathbf{v}; \mathcal{W}) \log\left[ \frac{P^{(1)}(\mathbf{v}; \mathcal{W})}{P(\mathbf{v}; \mathcal{W})} \right] \\
&= \left\langle \log\left[ \frac{P_t(\mathbf{v})}{P(\mathbf{v}; \mathcal{W})} \right] \right\rangle_{P_t(\mathbf{v})} - \left\langle \log\left[ \frac{P^{(1)}(\mathbf{v}; \mathcal{W})}{P(\mathbf{v}; \mathcal{W})} \right] \right\rangle_{P^{(1)}(\mathbf{v}; \mathcal{W})}.
\end{aligned}
$$

(3.34)

The brackets denote expectation values which can be approximated by samples distributed as $P_t(\mathbf{v})$ or $P^{(1)}(\mathbf{v}; \mathcal{W})$, respectively. To minimize the cost function we can apply the gradient descent algorithm. The derivative with respect to a general weight or bias $\mathcal{W}_\lambda$ is given by

$$
\frac{\partial C_{CD}(\mathcal{W})}{\partial \mathcal{W}_\lambda} = -\left\langle \frac{\partial \log\left[\tilde{P}(\mathbf{v}; \mathcal{W})\right]}{\partial \mathcal{W}_\lambda}\right\rangle_{P_t(\mathbf{v})} + \left\langle \frac{\partial \log\left[\tilde{P}(\mathbf{v}; \mathcal{W})\right]}{\partial \mathcal{W}_\lambda}\right\rangle_{P^{(1)}(\mathbf{v};\mathcal{W})} \quad (3.35)
$$
$$
- \frac{\partial P^{(1)}(\mathbf{v}; \mathcal{W})}{\partial \mathcal{W}_\lambda} \frac{\partial D_{KL}\left[P^{(1)}(\mathbf{v}; \mathcal{W}) \,\|\, P(\mathbf{v}; \mathcal{W})\right]}{\partial P^{(1)}(\mathbf{v}; \mathcal{W})},
$$

with unnormalized expressions $\tilde{P}(\mathbf{v}; \mathcal{W}) = Z(\mathcal{W})P(\mathbf{v}; \mathcal{W})$. Here we use

$$
\frac{\partial \log\left[P(\mathbf{v}; \mathcal{W})\right]}{\partial \mathcal{W}_\lambda} = \frac{\partial \log\left[\tilde{P}(\mathbf{v}; \mathcal{W})\right]}{\partial \mathcal{W}_\lambda} - \sum_{\{\tilde{\mathbf{v}}\}} P(\tilde{\mathbf{v}}; \mathcal{W}) \frac{\partial \log\left[\tilde{P}(\tilde{\mathbf{v}}; \mathcal{W})\right]}{\partial \mathcal{W}_\lambda}, \quad (3.36)
$$

and the fact that the terms resulting from the derivative of the normalization cancel. The last term in Eq. (3.35) is found to be small compared to the remaining terms, so that we can neglect it and get

$$
\frac{\partial C_{CD}(\mathcal{W})}{\partial \mathcal{W}_\lambda} \approx -\left\langle \frac{\partial \log\left[\tilde{P}(\mathbf{v}; \mathcal{W})\right]}{\partial \mathcal{W}_\lambda}\right\rangle_{P_t(\mathbf{v})} + \left\langle \frac{\partial \log\left[\tilde{P}(\mathbf{v}; \mathcal{W})\right]}{\partial \mathcal{W}_\lambda}\right\rangle_{P^{(1)}(\mathbf{v};\mathcal{W})}.
$$
$$
\qquad\qquad (3.37)
$$

The derivatives in Eq. (3.37) can be calculated explicitly for the RBM and yield for the case of connecting weights,

$$
\frac{\partial \log\left[\tilde{P}(\mathbf{v}; \mathcal{W})\right]}{\partial W_{k,l}} = \sum_{\{\mathbf{h}\}} \tilde{P}(\mathbf{h}|\,\mathbf{v}; \mathcal{W})\, v_k h_l, \quad (3.38)
$$

so that we can write the weight updates in the training procedure as

$$
W_{k,l}^{(\tau+1)} = W_{k,l}^{(\tau)} - \varepsilon \delta W_{k,l}, \quad (3.39)
$$
$$
\delta W_{k,l} = \frac{\partial C_{CD}(\mathcal{W})}{\partial W_{k,l}}
$$
$$
= -\langle v_k h_l\rangle_{P_t(\mathbf{v})P^{(1)}(\mathbf{h}|\mathbf{v};\mathcal{W})} + \langle v_k h_l\rangle_{P^{(1)}(\mathbf{v},\mathbf{h};\mathcal{W})}. \quad (3.40)
$$

Analogous expressions can be derived straightforwardly for the derivatives with respect to the biases. Here we can use the concept of Gibbs sampling, where the hidden variables are sampled from the conditional distribution $P(\mathbf{h}|\mathbf{v}; \mathcal{W})$. With this we can create the values of the hidden variable appearing in the first term of Eq. (3.40) by performing a single Gibbs sampling step on the given input data.

Thus, we get the following procedure to perform CD,

1. pick a sample from the given training data set and use it as a configuration of the visible variables $\mathbf{v}$,
2. sample a configuration of hidden variables $\mathbf{h}$ from $P(\mathbf{h}|\mathbf{v}; \mathcal{W})$ via a single Gibbs sampling step,
3. sample a configuration of visible variables $\tilde{\mathbf{v}}$ from $P(\tilde{\mathbf{v}}|\mathbf{h}; \mathcal{W})$ via a single Gibbs sampling step,
4. sample a configuration of hidden variables $\tilde{\mathbf{h}}$ from $P(\tilde{\mathbf{h}}|\tilde{\mathbf{v}}; \mathcal{W})$ via a single Gibbs sampling step,
5. calculate the weight updates,

$$
\begin{aligned}
\delta W_{k,l} &= -v_k h_l + \tilde{v}_k \tilde{h}_l, \\
\delta d_k &= -v_k + \tilde{v}_k, \\
\delta b_l &= -h_l + \tilde{h}_l, \\
\mathcal{W}_\lambda^{(\tau+1)} &= \mathcal{W}_\lambda^{(\tau)} - \varepsilon \delta \mathcal{W}_\lambda
\end{aligned}
\tag{3.41}
$$

In this procedure, one training sample is used at a time to calculate the weight update. A modification of this method would be to average over multiple training samples before calculating the weight update, which can make the training faster. On the other hand, it can also lead to cancellations in the averages, so that the learning rate needs to be adapted accordingly.

The ansatz introduced here is often referred to as $CD_1$, since only a single Gibbs sampling step is performed to see if the represented probability distribution stays close to the given one. For better accuracy, $k$ steps of Gibbs sampling can be performed here, which leads to the general method called $CD_k$, but this also makes the computations more expensive. The learning rate in the update scheme can be treated in the same way as in the supervised learning approach, so that a suitable value needs to be found for this hyper-parameter and also an adaptive ansatz might be helpful.

Besides RBM models, also ANN models called auto-encoders use unsupervised learning, which try to reproduce the given input in the output layer. However, the hidden layers in between have smaller numbers of variables, so that only the important features to reproduce the data are kept.

### 3.3.3 Reinforcement Learning

The third method to train an ANN is reinforcement learning, where an artificial software agent is considered acting in an environment and reacting on a reward he gets for several actions. This is for example used to train robots.

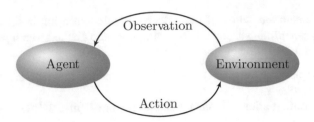

**Fig. 3.3** Schematic visualization of the setup used for reinforcement learning. An artificial software agent can perform actions in a fully or partially observed environment. It observes the environment and decides which action to perform next depending on the environment's state. The environment can then react on the agent's action by changing its state. For some actions or sequences of actions the agent gets a reward. The probability for the actions to be chosen given the environment's state is represented by an ANN which can be trained such that the reward gaining is maximized

The artificial agent in the reinforcement learning scheme can perform different actions in a fully or partially observed environment. The agent observes the state of the environment and based on this observation chooses its next action which can again affect the environment, as schematically visualized in Fig. 3.3. Here it is not known what the correct actions are, but after some actions or in some cases only after a sequence of actions, the agent gets a reward which can also be negative. The ANN is then trained such that the reward is maximized, for example in game playing the agent gets a positive reward at the end of the game if he won it and the probability to win is maximized.

As the reaction of the environment to the agent's actions is not directly known, we cannot optimize the final reward. Instead we can train the network by making the performed actions more likely if a high reward is received and making them less likely otherwise. In the case of getting the reward only after a sequence of actions, this can also make bad actions more likely, since the whole sequence gets a higher probability if a positive reward is obtained. However, taking many possible trajectories of actions into account, the bad ones are still suppressed in the sum. This training ansatz is known as policy gradient since the reward is in the end maximized using the stochastic gradient descent.

The probability for a certain trajectory $\tau = (\mathbf{a}, \mathbf{s})$ consisting of a sequence of actions $\mathbf{a}$ and states of the environment $\mathbf{s}$ at different times can then be written as

$$P(\tau; \mathcal{W}) = \prod_{t=1}^{t_{\text{end}}} P(s_{t+1}|\, a_t, s_t)\, P(a_t|\, s_t; \mathcal{W}). \qquad (3.42)$$

Here the product runs over all time steps. $P(s_{t+1}|s_t, a_t)$ is the probability of the environment to change to a state $s_{t+1}$ in the next time step given the actual state and action. Furthermore, $P(a_t|s_t; \mathcal{W})$ is the probability of the agent to perform the action $a_t$ at time $t$ given the environment is in state $s_t$. This probability can be represented by the ANN, as denoted by the sub-index $\mathcal{W}$, which is the set of all weights and biases in the network. These can be optimized, while the reaction of the environment is an ingredient to the network. The overall expected reward is then given by summing over all trajectories weighted according to their underlying distribution as stated in Eq. (3.42). The stochastic gradient descent can be applied to adapt the weights such that trajectories providing high rewards become more likely and others are suppressed. This way the reward is maximized during training.

## 3.4   Overview: Machine Learning in (Quantum) Physics

During the last years huge effort has been made in pointing out connections between artificial neural networks, or machine learning in general, and physics. In the following we give a brief overview of these connections, even though we can by far not make this overview complete here.

The basic and powerful difference between machine learning and other simulation methods is the statistical learning from examples instead of a logic-based approach. To better understand this statistical behavior, an analogy to statistical physics is used, for example given by considering neural networks as spin glasses [11–13]. Such glassy dynamics are useful when studying the complex energy landscapes of gradient descent methods in supervised learning schemes [14–17]. Statistical physics also helps to better understand unsupervised learning schemes such as contrastive divergence [6, 18].

Since there is a direct connection between the Boltzmann machine and the Ising model, the Boltzmann machine is often referred to as the "inverse Ising model". This connection is used to study the physical properties of the ANN, providing a better understanding [14, 19]. The training of the Boltzmann machine is additionally studied in the context of (non-)equilibrium and thermodynamic physics [18, 20–22], and a mean-field theory is used to further study deep Boltzmann machines [23].

While these are examples where one uses physics to gain a better understanding of machine learning methods, one can also use applications of machine learning methods to study physical problems. Considering statistical physics, machine learning is used to locate phase transitions [24–28] or to classify the corresponding phases [29–31]. It even finds applications in renormalization group theory [32].

Further applications of machine learning can be found in particle physics and cosmology, where one uses them for classification or regression tasks and moreover applies generative models [33, 34]. Here machine learning is applied for gravitational wave analysis [35] or black hole detection [36]. Considering the field of chemistry and materials, machine learning is used for data set generation or to analyze free-energy surfaces.

Another field where artificial neural networks find many applications is quantum physics [37, 38]. Here machine learning is used for quantum control [39, 40], to detect order in low-energy product states [41], or to find ground states [42]. Moreover, machine learning is applied to learn correlations in Einstein–Podolsky–Rosen (EPR) states [43], or to extract spatial geometry from entanglement [44]. Even the future trends of research in quantum physics are predicted via a neural-network-based scheme [45]. Furthermore, considering quantum many-body systems, a way to parametrize quantum states with neural networks has been developed and studied intensively. Especially a parametrization based on an RBM caught great attention. The representational power of this RBM-based parametrization has been studied in detail for wave functions of pure states as well as for mixed states [46–58]. Moreover, the effects of going to deeper networks in the parametrization, as well as different underlying network models have been analyzed [59–68].

Two main ansätze exist to learn the explicit representation of the states. One way is to use given numerical or experimental data and learn to represent it with an RBM, corresponding to quantum state tomography. Unsupervised [52, 69–71] or supervised [66, 67] learning schemes are used for this, where either only states with positive amplitudes are considered or different approaches are used to include the complex phases.

Another way to find the representation of states is a variational ansatz [48], which is mainly used to find ground states. Moreover, an extension to find first excited states has been introduced [61]. This variational learning scheme has been applied to find representations of wave functions for spin systems [48, 72–74] or bosonic [62, 66, 75] and fermionic [53, 76, 77] systems, and it has even been used for two-dimensional systems [48, 53, 54, 77]. Dynamics after sudden quenches can as well be represented via this variational learning ansatz [48, 78] and it finds applications when considering dissipative systems [55–58].

We further discuss and analyze this quantum state parametrization based on an RBM and the variational learning of the representation in this thesis. However, artificial neural networks find even more applications, such as classifying phases in quantum systems [24, 26, 63, 79, 80] or emulating basic quantum operations [81]. They are also used to improve existing simulation methods such as quantum Monte Carlo schemes [82–87]. Furthermore, they help in some cases to circumvent general problems, like the sign problem, in existing approximation methods [79, 88, 89].

The other way around, also ideas from quantum physics are used to inspire new machine learning methods. Tensor networks are considered as a basis for generative or discriminative learning [90–93] and matrix product states are used for classification tasks [94–96], or even in generative models [76, 97, 98]. When considering quantum computation, machine learning finds applications as well in controlling and preparing qubits [99, 100] or in error correction [101–105]. This provides help in building and studying quantum computers. The approach can again be turned around, as an acceleration in machine learning methods can be achieved by using classical or quantum hardware beyond the von-Neumann architecture. Examples for this are quantum computers [106, 107] or neuromorphic chips, which we further discuss in Sect. 3.5 [108–110].

While a lot of work has been done already on the connection between artificial neural networks and especially quantum physics, the field still leaves many open questions for further studies. This motivates us to analyze the parametrization of quantum states with an RBM and reachable benefits using support from neuromorphic hardware.

**Fig. 3.4** Duck-rabbit optical illusion. The interpretation of the seen picture switches between a rabbit (ears pointing to the left), or a duck (bill pointing to the left). This can be interpreted as drawing samples from a uniform distribution underlying the two interpretations and motivates performing statistical inference with spiking neural networks. Figure taken from [113]

## 3.5  Neuromorphic Hardware in the BrainScaleS Group

The BrainScaleS group at Heidelberg University is working on a neuromorphic hardware with physical models of neurons and synapses connecting them, which is implemented via mixed-signal microelectronics [109–112]. Their setup is based on so-called leaky integrate-and-fire (LIF) neurons with the neuron's membrane being modeled as a capacitance with an associated membrane potential. The potential is time-dependent and driven by inputs via synapses, the connections to other neurons. If the membrane potential crosses some threshold, the neuron emits an action potential, meaning its membrane shortly depolarizes (spikes) and by that sends a signal via the outgoing axon triggering the synapses of the connected downstream neurons [109, 110].

Considering a human brain, it happens that it has to deal with incomplete information. For example, when seeing Fig. 3.4 it is not clear whether a rabbit or a duck is shown. The brain needs to interpret the given information and decide what animal it sees. The interpretation can vary in time, so the status flips between seeing a duck or a rabbit. This behavior can be interpreted as drawing samples from a probability distribution. In the example both interpretations are equally probable. This observation led to the motivation of performing statistical inference via an ensemble of spiking neurons [109, 110, 114].

An implementation of Markov chain Monte Carlo sampling via spiking neurons has been derived in [114] and it has been translated to the LIF neurons in [110]. Its applicability has been benchmarked in [115]. In these implementations single binary random variables are represented by the single neurons. Via some temporal noise the network is driven such that it samples from a target probability distribution in the limit of large sampling times. The noise is implemented as high-frequency Poisson sources representing unmodeled parts of the brain, where a typical neuron has $10^4$ synaptic inputs.

The activity of a single neuron encodes the state of the corresponding random variable, where a spike of the neuron is interpreted as a switch of the binary variable from zero to one. Alternatively, this spike can also be interpreted as a switch from $-1$ to 1. This provides a conditional probability which can be used to define a Markov chain Monte Carlo sampling. As the membrane potential in the LIF neurons is deterministic, Poissonian noise can be added to make the sampling process stochastic [109, 110, 114–116].

Considering the membrane potential $u(t)$ of the LIF neurons, it can be approximately described by the following equation of motion [111],

$$\frac{du(t)}{dt} = \theta \left[ \mu - u(t) \right] + \sigma \eta(t).$$ (3.43)

Here $\theta$ is the strength of an attractive force pulling the potential to a mean value $\mu$ and $\sigma$ is given by the contribution of the Poissonian background. $\eta(t)$ is Gaussian random noise. The neuron emits a spike if the membrane potential $u(t)$ crosses a threshold $\vartheta$ from below and after the spike the membrane potential is set to a value $\rho$ for a refractory time $\tau_{\text{ref}}$, before it evolves further in time. Here only the actual membrane potential is affected. The effective potential given by the synaptic input is evolved further and can cause another spike right after the refractory time, as illustrated in Fig. 3.5 [109–111].

Equation (3.43) has the form of an Ornstein–Uhlenbeck process, a combination of a Wiener process and an attractive force term. The attractive force is given by the first term in Eq. (3.43). A Wiener process is a continuous generalization of a random walk and thus corresponds to the second term in Eq. (3.43). The probability distribution generated by an Ornstein–Uhlenbeck process is a unique solution of a Fokker–Planck equation, which is obeyed by the distribution of the synaptic input of LIF neurons [109–111].

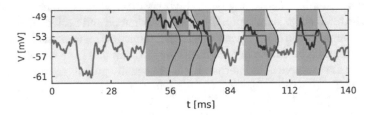

**Fig. 3.5** Dynamical evolution of the membrane potential found for a spiking LIF neuron. The effective potential $u(t)$ is shown by the red line, while the actual membrane potential is depicted in blue, showing the reset to the potential $\rho$ for a refractory time $\tau_{\text{ref}}$ after each spike, indicated by the gray regions. A spike is emitted if the threshold $\vartheta$ depicted by the black line is crossed from below. If the effective potential is still above the threshold after the refractory time, the neuron spikes again, yielding a burst of spikes. The pink areas denote the transition probability $P(u_{i+1} | u_i)$, illustrating a prediction for the effective membrane potential. Figure taken from [109]

This correspondence between the neuron dynamics and the Ornstein–Uhlenbeck process can be used to perform stochastic inference. To get the probability $P(z = 1)$ of a neuron $z$ to be in an active state, the burst lengths and the mean first passage times can be considered [109–111]. Burst lengths are defined as the numbers of spikes a neuron performs in a row, so it spikes again directly after the refractory time. On the other hand, mean first passage times are defined by the mean duration until the threshold is crossed again from below after a spike. Evaluating these quantities for the Ornstein–Uhlenbeck process yields [109–111]

$$P(z = 1) = \frac{\sum_n P_n n \tau_{\text{ref}}}{\sum_n P_n \left(n \tau_{\text{ref}} + \sum_{k=1}^{n-1} \bar{\tau}_k^b + T_n\right)}, \tag{3.44}$$

where the sum runs over the burst lengths $n$ with underlying distribution $P_n$. $T_n$ is the distribution underlying the mean first passage times, and $\bar{\tau}_k^b$ is the mean drift time to cross the threshold when starting from the potential $\rho$ after the refractory time [109–111].

A neuron in a high conductance state is defined as experiencing strong synaptic stimulation and thus shows accelerated membrane dynamics, symmetrizing the neural activation function. For such neurons, the probability of being active can be approximated with good accuracy by a sigmoid function,

$$P(z = 1 | u) \approx \frac{1}{1 + \exp\left[-\frac{u - u_0}{\alpha}\right]}, \tag{3.45}$$

with neuron potential $u$, a relative offset $u_0$ and a rescaling factor $\alpha$ ensuring the right units. This corresponds to the activation function used in Markov chain Monte Carlo sampling from Boltzmann distributions and can be generalized to a network of LIF neurons in the high conductance state. It has been shown that such a network can be configured to sample from a Boltzmann distribution [109–111, 115].

To train the weights in the implemented neural network such that they provide a desired Boltzmann distribution on which the sampling can be performed, the contrastive divergence scheme can be used, see Sect. 3.3.2, which runs on an external computer. An on-chip implementation based on Spike Time Dependent Plasticity could be used on future hardware platforms [115].

As BrainScaleS implements the sampling network in analog circuits, its runtime is only dependent on the time constants of the model. Using the typical parameters from [109, 110, 115] this means that a new sample is generated every micro second, independent of the system size that is emulated. This leads to an expected speedup of at least one order of magnitude in the sample creation compared to simulations on a classical computer, while using about three orders of magnitude less energy [109, 117].

# References

1. Bishop C (2006) Pattern recognition and machine learning. Springer, New York. https://www. springer.com/de/book/9780387310732?referer=www.springer.de
2. Marquardt F (2017) Machine learning for physicists. https://machine-learning-for-physicists. org/
3. Mehta P, Bukov M, Wang C-H, Day AGR, Richardson C, Fisher CK, Schwab DJ (2019) A high-bias, low-variance introduction to machine learning for physicists. Phys Rep 810:1–124. http://www.sciencedirect.com/science/article/pii/S0370157319300766
4. Hinton GE (2012) A practical guide to training restricted boltzmann machines, pp 599–619. Springer, Berlin. https://doi.org/10.1007/978-3-642-35289-8_32
5. Nielsen MA (2015) Neural networks and deep learning. Determination Press. http:// neuralnetworksanddeeplearning.com/
6. Hinton GE (2002) Training products of experts by minimizing contrastive divergence. Neural Comput 14(8):1771–1800. https://doi.org/10.1162/089976602760128018
7. Carleo G, Cirac I, Cranmer K, Daudet L, Schuld M, Tishby N, Vogt-Maranto L, Zdeborová L (2019) Machine learning and the physical sciences. Rev Mod Phys 91:045002. https://link. aps.org/doi/10.1103/RevModPhys.91.045002
8. Silver D, Huang A, Maddison CJ, Guez A, Sifre L, van den Driessche G, Schrittwieser J, Antonoglou I, Panneershelvam V, Lanctot M, Dieleman S, Grewe D, Nham J, Kalchbrenner N, Sutskever I, Lillicrap T, Leach M, Kavukcuoglu K, Graepel T, Hassabis D (2016) Mastering the game of Go with deep neural networks and tree search. Nature 529:484. https://doi.org/ 10.1038/nature16961
9. Montúfar G, Morton J (2015) Discrete restricted Boltzmann machines. J Mach Learn Res 16:653–672. http://jmlr.org/papers/v16/montufar15a.html
10. Le Roux N, Bengio Y (2008) Representational power of restricted Boltzmann machines and deep belief networks. Neural Comput 20(6):1631–1649. https://doi.org/10.1162/neco.2008. 04-07-510
11. Valiant LG (1984) A theory of the learnable. Commun ACM 27(11):1134–1142. http://doi. acm.org/10.1145/1968.1972
12. Hopfield JJ (1982) Neural networks and physical systems with emergent collective computational abilities. PNAS 79(8):2554–2558. https://www.pnas.org/content/79/8/2554
13. Amit DJ, Gutfreund H, Sompolinsky H (1985) Spin-glass models of neural networks. Phys Rev A 32:1007–1018. https://link.aps.org/doi/10.1103/PhysRevA.32.1007
14. Baity-Jesi M, Sagun L, Geiger M, Spigler S, Arous GB, Cammarota C, LeCun Y, Wyart M, Biroli G (2018) Comparing dynamics: deep neural networks versus glassy systems, Proceedings of machine learning research, vol 80, pp 314–323. PMLR, Stockholmsmässan, Stockholm Sweden. Accessed 10–15 Jul 2018. http://proceedings.mlr.press/v80/baity-jesi18a.html
15. Baldassi C, Borgs C, Chayes JT, Ingrosso A, Lucibello C, Saglietti L, Zecchina R (2016) Unreasonable effectiveness of learning neural networks: from accessible states and robust ensembles to basic algorithmic schemes. PNAS 113(48):E7655–E7662. https://www.pnas. org/content/113/48/E7655
16. Baldassi C, Ingrosso A, Lucibello C, Saglietti L, Zecchina R (2015) Subdominant dense clusters allow for simple learning and high computational performance in neural networks with discrete synapses. Phys Rev Lett 115:128101. https://link.aps.org/doi/10.1103/PhysRevLett. 115.128101
17. Sidky H, Whitmer JK (2018) Learning free energy landscapes using artificial neural networks. J Chem Phys 148(10):104111. https://doi.org/10.1063/1.5018708
18. Decelle A, Fissore G, Furtlehner C (2017) Spectral dynamics of learning in restricted Boltzmann machines. EPL 119(6):60001. https://doi.org/10.1209%2F0295-5075%2F119 %2F60001
19. Chau Nguyen H, Zecchina R, Berg J (2017) Inverse statistical problems: from the inverse Ising problem to data science. Adv Phys 66(3):197–261. https://doi.org/10.1080/00018732. 2017.1341604

20. Cossu G, Del Debbio L, Giani T, Khamseh A, Wilson M (2019) Machine learning determination of dynamical parameters: the ising model case. Phys Rev B 100:064304. https://link.aps.org/doi/10.1103/PhysRevB.100.064304
21. Funai SS, Giataganas D (2018) Thermodynamics and feature extraction by machine learning. arXiv:1810.08179 [cond-mat.stat-mech]
22. Salazar DSP (2017) Nonequilibrium thermodynamics of restricted Boltzmann machines. Phys Rev E 96:022131. https://link.aps.org/doi/10.1103/PhysRevE.96.022131
23. Huang H (2018) Mechanisms of dimensionality reduction and decorrelation in deep neural networks. Phys Rev E 98:062313. https://link.aps.org/doi/10.1103/PhysRevE.98.062313
24. Carrasquilla J, Melko RG (2017) Machine learning phases of matter. Nat Phys 13:431. http://dx.doi.org/10.1038/nphys4035
25. Morningstar A, Melko RG (2018) Deep learning the Ising model near criticality. J Mach Learn Res 18(163):1–17. http://jmlr.org/papers/v18/17-527.html
26. van Nieuwenburg EPL, Liu Y-H, Huber SD (2017) Learning phase transitions by confusion. Nat Phys 13:435. http://dx.doi.org/10.1038/nphys4037
27. Wetzel SJ (2017) Unsupervised learning of phase transitions: from principal component analysis to variational autoencoders. Phys Rev E 96:022140. https://link.aps.org/doi/10.1103/PhysRevE.96.022140
28. Rządkowski W, Defenu N, Chiacchiera S, Trombettoni A, Bighin G (2019) Detecting hidden and composite orders in layered models via machine learning. arXiv:1907.05417 [cond-mat.dis-nn]
29. Wetzel SJ, Scherzer M (2017) Machine learning of explicit order parameters: from the Ising model to SU(2) lattice gauge theory. Phys Rev B 96:184410. https://link.aps.org/doi/10.1103/PhysRevB.96.184410
30. Wang C, Zhai H (2017) Machine learning of frustrated classical spin models I: principal component analysis. Phys Rev B 96:144432. https://doi.org/10.1007/s11467-018-0798-7
31. Wang C, Zhai H (2018) Machine learning of frustrated classical spin models (II): Kernel principal component analysis. Front Phys 13(5):130507. https://doi.org/10.1007/s11467-018-0798-7
32. Mehta P, Schwab DJ (2014) An exact mapping between the variational renormalization group and deep learning. arXiv:1410.3831 [stat.ML]
33. Guest D, Cranmer K, Whiteson D (2018) Deep learning and its application to LHC physics. Ann Rev Nucl Part S 68(1):161–181. https://doi.org/10.1146/annurev-nucl-101917-021019
34. Radovic A, Williams M, Rousseau D, Kagan M, Bonacorsi D, Himmel A, Aurisano A, Terao K, Wongjirad T (2018) Machine learning at the energy and intensity frontiers of particle physics. Nature 560(7716):41–48. https://doi.org/10.1038/s41586-018-0361-2
35. Biswas R, Blackburn L, Cao J, Essick R, Hodge KA, Katsavounidis E, Kim K, Kim Y-M, Le Bigot E, Lee C-H, Oh JJ, Oh SH, Son EJ, Tao Y, Vaulin R, Wang X (2013) Application of machine learning algorithms to the study of noise artifacts in gravitational-wave data. Phys Rev D 88:062003. https://link.aps.org/doi/10.1103/PhysRevD.88.062003
36. Pasquato M (2016) Detecting intermediate mass black holes in globular clusters with machine learning. arXiv:1606.08548 [astro-ph.GA]
37. Biamonte J, Wittek P, Pancotti N, Rebentrost P, Wiebe N, Lloyd S (2017) Quantum machine learning. Nature 549:195. http://dx.doi.org/10.1038/nature23474
38. Dunjko V, Briegel HJ (2017) Machine learning & artificial intelligence in the quantum domain. arXiv:1709.02779 [quant-ph]
39. Zahedinejad E, Ghosh J, Sanders BC (2016) Designing high-fidelity single-shot three-qubit gates: a machine-learning approach. Phys Rev Appl 6:054005. https://link.aps.org/doi/10.1103/PhysRevApplied.6.054005
40. August M, Ni X (2017) Using recurrent neural networks to optimize dynamical decoupling for quantum memory. Phys Rev A 95:012335. https://link.aps.org/doi/10.1103/PhysRevA.95.012335
41. Rao W-J, Li Z, Zhu Q, Luo M, Wan X (2018) Identifying product order with restricted Boltzmann machines. Phys Rev B 97:094207. https://link.aps.org/doi/10.1103/PhysRevB.97.094207

42. Wu J, Zhang W (2019) Finding quantum many-body ground states with artificial neural network. arXiv:1906.11216 [cond-mat.dis-nn]
43. Weinstein S (2017) Learning the Einstein-Podolsky-Rosen correlations on a restricted Boltzmann machine. arXiv:1707.03114 [quant-ph]
44. You Y-Z, Yang Z, Qi X-L (2018) Machine learning spatial geometry from entanglement features. Phys Rev B 97:045153. https://link.aps.org/doi/10.1103/PhysRevB.97.045153
45. Krenn M, Zeilinger A (2020) Predicting research trends with semantic and neural networks with an application in quantum physics. Proc Natl Acad Sci 117(4):1910–1916. https://www.pnas.org/content/117/4/1910
46. Czischek S, Gärttner M, Gasenzer T (2018) Quenches near Ising quantum criticality as a challenge for artificial neural networks. Phys Rev B 98:024311. https://doi.org/10.1103/PhysRevB.98.024311
47. Torlai G, Melko RG (2016) Learning thermodynamics with Boltzmann machines. Phys Rev B 94:165134. https://link.aps.org/doi/10.1103/PhysRevB.94.165134
48. Carleo G, Troyer M (2017) Solving the quantum many-body problem with artificial neural networks. Science 355(6325):602–606. http://science.sciencemag.org/content/355/6325/602
49. Lu S, Gao X, Duan L-M (2019) Efficient representation of topologically ordered states with restricted Boltzmann machines. Phys Rev B 99:155136. https://link.aps.org/doi/10.1103/PhysRevB.99.155136
50. Gao X, Duan L-M (2017) Efficient representation of quantum many-body states with deep neural networks. Nat Commun 8(1). https://doi.org/10.1038/s41467-017-00705-2
51. Deng D-L, Li X, Das Sarma S (2017) Quantum entanglement in neural network states. Phys Rev X 7:021021. https://link.aps.org/doi/10.1103/PhysRevX.7.021021
52. Carrasquilla J, Torlai G, Melko RG, Aolita L (2019) Reconstructing quantum states with generative models. Nat Mach Intell 1(3):155–161. https://doi.org/10.1038/s42256-019-0028-1
53. Nomura Y, Darmawan AS, Yamaji Y, Imada M (2017) Restricted Boltzmann machine learning for solving strongly correlated quantum systems. Phys Rev B 96:205152. https://link.aps.org/doi/10.1103/PhysRevB.96.205152
54. Westerhout T, Astrakhantsev N, Tikhonov KS, Katsnelson M, Bagrov AA (2020) Generalization properties of neural network approximations to frustrated magnet ground states. Nat Commun 11(1). https://doi.org/10.1038/s41467-020-15402-w
55. Hartmann MJ, Carleo G (2019) Neural-network approach to dissipative quantum many-body dynamics. Phys Rev Lett 122:250502. https://link.aps.org/doi/10.1103/PhysRevLett.122.250502
56. Nagy A, Savona V (2019) Variational quantum Monte Carlo method with a neural-network ansatz for open quantum systems. Phys Rev Lett 122:250501. https://link.aps.org/doi/10.1103/PhysRevLett.122.250501
57. Vicentini F, Biella A, Regnault N, Ciuti C (2019) Variational neural-network ansatz for steady states in open quantum systems. Phys Rev Lett 122:250503. https://link.aps.org/doi/10.1103/PhysRevLett.122.250503
58. Yoshioka N, Hamazaki R (2019) Constructing neural stationary states for open quantum many-body systems. Phys Rev B 99:214306. https://link.aps.org/doi/10.1103/PhysRevB.99.214306
59. Carleo G, Nomura Y, Imada M (2018) Constructing exact representations of quantum many-body systems with deep neural networks. Nat Commun 9(1). https://doi.org/10.1038/s41467-018-07520-3
60. Freitas N, Morigi G, Dunjko V (2018) Neural network operations and Susuki-Trotter evolution of neural network states. Int J Quantum Inf 16(08):1840008. https://doi.org/10.1142/S0219749918400087
61. Choo K, Carleo G, Regnault N, Neupert T (2018) Symmetries and many-body excitations with neural-network quantum states. Phys Rev Lett 121:167204. https://link.aps.org/doi/10.1103/PhysRevLett.121.167204

62. Saito H (2018) Method to solve quantum few-body problems with artificial neural networks. J Phys Soc Jpn 87(7):074002. https://doi.org/10.7566/JPSJ.87.074002
63. Sharir O, Levine Y, Wies N, Carleo G, Shashua A (2020) Deep autoregressive models for the efficient variational simulation of many-body quantum systems. Phys Rev Lett 124:020503. https://link.aps.org/doi/10.1103/PhysRevLett.124.020503
64. Levine Y, Sharir O, Cohen N, Shashua A (2019) Quantum entanglement in deep learning architectures. Phys Rev Lett 122:065301. https://link.aps.org/doi/10.1103/PhysRevLett.122.065301
65. Liu D, Ran S-J, Wittek P, Peng C, García RB, Su G, Lewenstein M (2019) Machine learning by unitary tensor network of hierarchical tree structure. New J Phys 21(7):073059. https://doi.org/10.1088%2F1367-2630%2Fab31ef
66. Saito H (2017) Solving the Bose-Hubbard model with machine learning. J Phys Soc Jpn 86(9):093001. https://doi.org/10.7566/JPSJ.86.093001
67. Cai Z, Liu J (2018) Approximating quantum many-body wave functions using artificial neural networks. Phys Rev B 97:035116. https://link.aps.org/doi/10.1103/PhysRevB.97.035116
68. Levine Y, Yakira D, Cohen N, Shashua A (2017) Deep learning and quantum entanglement: fundamental connections with implications to network design. arXiv:1704.01552 [cs.LG]
69. Torlai G, Mazzola G, Carrasquilla J, Troyer M, Melko R, Carleo G (2018) Neural-network quantum state tomography. Nat Phys 14:447–450. https://doi.org/10.1038/s41567-018-0048-5
70. Torlai G, Melko RG (2018) Latent space purification via neural density operators. Phys Rev Lett 120:240503. https://link.aps.org/doi/10.1103/PhysRevLett.120.240503
71. Torlai G, Timar B, van Nieuwenburg EPL, Levine H, Omran A, Keesling A, Bernien H, Greiner M, Vuletić V, Lukin MD, Melko RG, Endres M (2019) Integrating neural networks with a quantum simulator for state reconstruction. Phys Rev Lett 123:230504. https://link.aps.org/doi/10.1103/PhysRevLett.123.230504
72. Deng D-L, Li X, Das Sarma S (2017) Machine learning topological states. Phys Rev B 96:195145. https://link.aps.org/doi/10.1103/PhysRevB.96.195145
73. Glasser I, Pancotti N, August M, Rodriguez ID, Cirac JI (2018) Neural-network quantum states, string-bond states, and chiral topological states. Phys Rev X 8:011006. https://link.aps.org/doi/10.1103/PhysRevX.8.011006
74. Liang X, Liu W-Y, Lin P-Z, Guo G-C, Zhang Y-S, He L (2018) Solving frustrated quantum many-particle models with convolutional neural networks. Phys Rev B 98:104426. https://link.aps.org/doi/10.1103/PhysRevB.98.104426
75. Saito H, Kato M (2018) Machine learning technique to find quantum many-body ground states of bosons on a lattice. J Phys Soc Jpn 87(1):014001. https://doi.org/10.7566/JPSJ.87.014001
76. Han J, Zhang L, Weinan E (2018) Solving many-electron Schrödinger equation using deep neural networks. arXiv:1807.07014 [physics.comp-ph]
77. Luo D, Clark BK (2019) Backflow transformations via neural networks for quantum many-body wave functions. Phys Rev Lett 122:226401. https://link.aps.org/doi/10.1103/PhysRevLett.122.226401
78. Schmitt M, Heyl M (2018) Quantum dynamics in transverse-field Ising models from classical networks. SciPost Phys 4:013. https://scipost.org/10.21468/SciPostPhys.4.2.013
79. Broecker P, Carrasquilla J, Melko RG, Trebst S (2017) Machine learning quantum phases of matter beyond the fermion sign problem. Sci Rep 7:8823. https://doi.org/10.1038/s41598-017-09098-0
80. Wang L (2016) Discovering phase transitions with unsupervised learning. Phys Rev B 94:195105. https://link.aps.org/doi/10.1103/PhysRevB.94.195105
81. Pehle C, Meier K, Oberthaler M, Wetterich C (2018) Emulating quantum computation with artificial neural networks. arXiv:1810.10335 [quant-ph]
82. Shen H, Liu J, Liang F (2018) Self-learning Monte Carlo with deep neural networks. Phys Rev B 97:205140. https://link.aps.org/doi/10.1103/PhysRevB.97.205140
83. Liu J, Shen H, Qi Y, Meng ZY, Fu L (2017) Self-learning Monte Carlo method and cumulative update in fermion systems. Phys Rev B 95:241104. https://link.aps.org/doi/10.1103/PhysRevB.95.241104

84. Nagai Y, Shen H, Qi Y, Liu J, Liang F (2017) Self-learning Monte Carlo method: continuous-time algorithm. Phys Rev B 96:161102. https://link.aps.org/doi/10.1103/PhysRevB.96.161102

85. Liu J, Qi Y, Meng ZY, Fu L (2017) Self-learning Monte Carlo method. Phys Rev B 95:041101. https://link.aps.org/doi/10.1103/PhysRevB.95.041101

86. Inack EM, Santoro GE, Dell'Anna L, Pilati S (2018) Projective quantum Monte Carlo simulations guided by unrestricted neural network states. Phys Rev B 98:235145. https://link.aps.org/doi/10.1103/PhysRevB.98.235145

87. Pilati S, Inack EM, Pieri P (2019) Self-learning projective quantum monte carlo simulations guided by restricted boltzmann machines. Phys Rev E 100:043301. https://link.aps.org/doi/10.1103/PhysRevE.100.043301

88. Torlai G, Carrasquilla J, Fishman MT, Melko RG, Fisher MPA (2019) Wavefunction positivization via automatic differentiation. arXiv:1906.04654 [quant-ph]

89. Hangleiter D, Roth I, Nagaj D, Eisert J (2019) Easing the Monte Carlo sign problem. arXiv:1906.02309 [quant-ph]

90. Huang Y, Moore JE (2017) Neural network representation of tensor network and chiral states. arXiv:1701.06246 [cond-mat.dis-nn]

91. Huggins W, Patil P, Mitchell B, Whaley KB, Stoudenmire EM (2019) Towards quantum machine learning with tensor networks. Quantum Sci Technol 4(2):024001. https://doi.org/10.1088%2F2058-9565%2Faaea94

92. Chen J, Cheng S, Xie H, Wang L, Xiang T (2018) Equivalence of restricted Boltzmann machines and tensor network states. Phys Rev B 97:085104. https://link.aps.org/doi/10.1103/PhysRevB.97.085104

93. Efthymiou S, Hidary J, Leichenauer S (2019) TensorNetwork for machine learning. arXiv:1906.06329 [cs.LG]

94. Liu Y, Zhang X, Lewenstein M, Ran S-J (2018) Entanglement-guided architectures of machine learning by quantum tensor network. arXiv:1803.09111 [stat.ML]

95. Novikov A, Trofimov M, Oseledets I (2016) Exponential machines. arXiv:1605.03795 [stat.ML]

96. Stoudenmire E, Schwab DJ (2016) Supervised learning with tensor networks, pp. 4799–4807. Curran Associates, Inc. http://papers.nips.cc/paper/6211-supervised-learning-with-tensor-networks.pdf

97. Han Z-Y, Wang J, Fan H, Wang L, Zhang P (2018) Unsupervised generative modeling using matrix product states. Phys Rev X 8:031012. https://link.aps.org/doi/10.1103/PhysRevX.8.031012

98. Stokes J, Terilla J (2019) Probabilistic modeling with matrix product states. Entropy 21(12):1236. http://dx.doi.org/10.3390/e21121236

99. Bukov M, Day AGR, Sels D, Weinberg P, Polkovnikov A, Mehta P (2018) Reinforcement learning in different phases of quantum control. Phys Rev X 8:031086. https://link.aps.org/doi/10.1103/PhysRevX.8.031086

100. Bukov M (2018) Reinforcement learning for autonomous preparation of Floquet-engineered states: Inverting the quantum Kapitza oscillator. Phys Rev B 98:224305. https://link.aps.org/doi/10.1103/PhysRevB.98.224305

101. Torlai G, Melko RG (2017) Neural decoder for topological codes. Phys Rev Lett 119:030501. https://link.aps.org/doi/10.1103/PhysRevLett.119.030501

102. Krastanov S, Jiang L (2017) Deep neural network probabilistic decoder for stabilizer codes. Sci Rep 7(11003). https://doi.org/10.1038/s41598-017-11266-1

103. Varsamopoulos S, Criger B, Bertels K (2017) Decoding small surface codes with feedforward neural networks. Quantum Sci Technol 3(1):015004. https://doi.org/10.1088%2F2058-9565%2Faa955a

104. Baireuther P, O'Brien TE, Tarasinski B, Beenakker CWJ (2018) Machine-learning-assisted correction of correlated qubit errors in a topological code. Quantum 2:48. https://doi.org/10.22331/q-2018-01-29-48

105. Sweke R, Kesselring MS, van Nieuwenburg EPL, Eisert J (2018) Reinforcement learning decoders for fault-tolerant quantum computation. arXiv:1810.07207 [quant-ph]
106. Torlai G, Melko RG (2019) Machine learning quantum states in the NISQ era. arXiv:1905.04312 [quant-ph]
107. Gardas B, Rams MM, Dziarmaga J (2018) Quantum neural networks to simulate many-body quantum systems. Phys Rev B 98:184304. https://link.aps.org/doi/10.1103/PhysRevB.98.184304
108. Di Ventra M, Traversa FL (2018) Perspective: memcomputing: Leveraging memory and physics to compute efficiently. J Appl Phys 123(18):180901. https://doi.org/10.1063/1.5026506
109. Petrovici MA (2016) Form versus function: theory and models for neuronal substrates. Springer International Publishing, Berlin. https://doi.org/10.1007/978-3-319-39552-4
110. Petrovici MA, Bill J, Bytschok I, Schemmel J, Meier K (2016) Stochastic inference with spiking neurons in the high-conductance state. Phys Rev E 94:042312. https://link.aps.org/doi/10.1103/PhysRevE.94.042312
111. Kades RG, Pawlowski J (2019)The discrete Langevin machine: Bridging the gap between thermodynamic and neuromorphic systems. arxiv:1901.05214 [cs.NE]
112. Schemmel J, Brüderle D, Grübl A, Hock M, Meier K, Millner S (2010) A wafer-scale neuro-morphic hardware system for large-scale neural modeling, pp 1947–1950. https://ieeexplore.ieee.org/document/5536970/
113. Jastrow J (1900) Fact and fable in psychology. Houghton, Mifflin and Company, Boston. https://books.google.de/books?id=xPiiv3SCOacC
114. Buesing L, Bill J, Nessler B, Maass W (2011) Neural dynamics as sampling: a model for stochastic computation in recurrent networks of spiking neurons. PLoS Comput Biol 7(11):1–22, 11. https://doi.org/10.1371/journal.pcbi.1002211
115. Kungl AF, Schmitt S, Klähn J, Müller P, Baumbach A, Dold D, Kugele A, Müller E, Koke C, Kleider M, Mauch C, Breitwieser O, Leng L, Gürtler N, Güttler M, Husmann D, Husmann K, Hartel A, Karasenko V, Grübl A, Schemmel J, Meier K, Petrovici MA (2019) Accelerated physical emulation of bayesian inference in spiking neural networks. Front Neurosci 13:1201. https://www.frontiersin.org/article/10.3389/fnins.2019.01201
116. Stochasticity from function — why the bayesian brain may need no noise. Neur Netw 119:200–213 (2019). https://doi.org/10.1016/j.neunet.2019.08.002
117. Wunderlich T, Kungl AF, Müller E, Hartel A, Stradmann Y, Aamir SA, Grübl A, Heimbrecht A, Schreiber K, Stöckel D, Pehle C, Billaudelle S, Kiene G, Mauch C, Schemmel J, Meier K, Petrovici MA (2019) Demonstrating advantages of neuromorphic computation: a pilot study. Front Neurosci 13:260. https://doi.org/10.3389/fnins.2019.00260

# Part II
# Simulations of Quantum Many-Body Systems

# Chapter 4
# Discrete Truncated Wigner Approximation

One approach to approximate dynamics of quantum systems is to consider the classical time evolution of initial configurations which are sampled from the Wigner function, a (quasi-)probability distribution in phase space underlying the initial quantum state. This ansatz is called the truncated Wigner approximation and it can be adapted for spin-1/2 systems providing the discrete truncated Wigner approximation. In comparison to mean-field approximations, this simulation method includes quantum fluctuations resulting from sampling the initial configuration and averaging over the outcome of many classical trajectories calculated from different initial states.

We introduce the general truncated Wigner approximation in Sect. 4.1 based on detailed discussions in [1, 2]. Afterwards, we introduce the modifications to obtain the discrete truncated Wigner approximation and benchmark this method on sudden quenches in the transverse-field Ising model.

The chapter contains discussions and results based on [3].

## 4.1 Truncated Wigner Approximation

In contrast to classical mechanics, according to Heisenberg's uncertainty principle there are always fluctuations present in quantum mechanical systems. Thus, the state of a quantum system can never be fully determined, e.g., position and momentum cannot be measured exactly at the same time. These so-called quantum fluctuations are included in the density-operator representation of the states. The expectation value of a general operator $O$ can be calculated as stated in Eq. (2.7),

S. Czischek, *Neural-Network Simulation of Strongly Correlated Quantum Systems*,
Springer Theses, https://doi.org/10.1007/978-3-030-52715-0_4

$$\langle O \rangle = \sum_{\{\mathbf{v}\}} \sum_{\{\tilde{\mathbf{v}}\}} \langle \tilde{\mathbf{v}} | O | \mathbf{v} \rangle c_{\tilde{\mathbf{v}}}^* c_{\mathbf{v}} \tag{4.1}$$

$$= \mathrm{Tr}\left[ O \rho \right],$$

$$\rho = \sum_{\{\mathbf{v}\}} \sum_{\{\tilde{\mathbf{v}}\}} |\mathbf{v} \rangle \langle \tilde{\mathbf{v}} | c_{\tilde{\mathbf{v}}}^* c_{\mathbf{v}}. \tag{4.2}$$

We express the density operator $\rho$ in terms of the basis states weighted by the amplitudes $c_{\tilde{\mathbf{v}}}^* c_{\mathbf{v}}$. The form of Eq. (4.1) shows a direct connection to calculating expectation values in statistical physics via an integral over a variable, weighted by the underlying probability distribution. This gives the motivation to consider quantum mechanics in a statistical language [4–6].

Quantum systems can be represented in terms of quantum phase-space variables, which leads to an efficient way of dealing with quantum fluctuations in a statistical language. The quantum phase space can be used to describe quantum mechanics analogously to the classical phase space in classical physics. Hence, a general quantum operator corresponds to a classical function of phase-space variables [7]. Quantum fluctuations can then be captured by statistical fluctuations, where the underlying probability distribution is given by the expression of the density matrix in this phase-space language. This is, however, not necessarily a positive definite function. In the phase-space representation dynamics near the classical limit, in regimes of small quantum fluctuations, can also be calculated efficiently, leading to the semi-classical truncated Wigner approximation (TWA).

Analogously to the phase space of classical systems, the quantum phase space can be represented in the position-momentum basis. It is also possible to span it by coherent states as eigenstates of the annihilation operator, and all of the following calculations can also be applied for this case. For convenience sake we stick to the example of the position-momentum basis in the following.

For the remainder of this section we denote quantum operators using hats to distinguish them from classical variables. To map a general quantum operator $\hat{O}$ uniquely into the phase space, we introduce its Weyl symbol $O_W$, which is a classical function in phase space. A one-to-one mapping can be defined via

$$O_{\mathrm{W}}\left(\mathbf{x}, \mathbf{p}\right) = \int \mathrm{d}^D \boldsymbol{\xi} \left\langle \mathbf{x} - \frac{\boldsymbol{\xi}}{2} \middle| \hat{O}\left(\hat{\mathbf{x}}, \hat{\mathbf{p}}\right) \middle| \mathbf{x} + \frac{\boldsymbol{\xi}}{2} \right\rangle \exp\left[ \frac{i}{\hbar} \mathbf{p} \cdot \boldsymbol{\xi} \right]. \tag{4.3}$$

The position coefficients of the phase space are here given by $\mathbf{x}$, where $\hat{\mathbf{x}}$ is the position operator. Analogously, $\mathbf{p}$ is the momentum coefficient with momentum operator $\hat{\mathbf{p}}$. The vectors are $D$-dimensional, with $D = Nd$ for $N$-particle systems in $d$ spatial dimensions. The phase space has then dimension $2D$.

One special Weyl symbol is the Wigner function, $W(\mathbf{x}, \mathbf{p})$, which results from mapping the density matrix into phase space [8],

$$W(\mathbf{x}, \mathbf{p}) := \rho_{\mathrm{W}}(\mathbf{x}, \mathbf{p})$$

$$= \int d^{D}\boldsymbol{\xi} \left\langle \mathbf{x} - \frac{\boldsymbol{\xi}}{2} \middle| \hat{\rho} \middle| \mathbf{x} + \frac{\boldsymbol{\xi}}{2} \right\rangle \exp\left[\frac{i}{\hbar}\mathbf{p}\cdot\boldsymbol{\xi}\right]. \tag{4.4}$$

It shows properties of a probability distribution, specifically it is normalized,

$$\int \int \frac{d^{D}\mathbf{x}\, d^{D}\mathbf{p}}{(2\pi\hbar)^{D}} W(\mathbf{x}, \mathbf{p}) = 1, \tag{4.5}$$

but it can take negative values and is therefore called a quasi-probability distribution. The Wigner function can be used as the weighting distribution when calculating expectation values of Weyl symbols, since it results from the density matrix and thus includes the quantum fluctuations. The definition of the Weyl symbols and the Wigner function provides a complete description of a quantum state, so that expectation values of operators can be calculated via

$$\langle \hat{O}(\hat{\mathbf{x}}, \hat{\mathbf{p}}) \rangle = \int \int \frac{d^{D}\mathbf{x}\, d^{D}\mathbf{p}}{(2\pi\hbar)^{D}} W(\mathbf{x}, \mathbf{p})\, O_{\mathrm{W}}(\mathbf{x}, \mathbf{p}). \tag{4.6}$$

For many important classes of quantum states the Wigner function is even positive or can at least be approximated by a positive function. In these cases a Monte Carlo scheme can be used to draw $Q$ samples $(\mathbf{x}_q, \mathbf{p}_q)$ from the Wigner function and approximate the expectation value of a quantum operator,

$$\langle \hat{O}(\hat{\mathbf{x}}, \hat{\mathbf{p}}) \rangle \approx \frac{1}{Q}\sum_{q=1}^{Q} O_{\mathrm{W}}(\mathbf{x}_q, \mathbf{p}_q)$$

$$= \langle O_{\mathrm{W}}(\mathbf{x}, \mathbf{p}) \rangle_{W(\mathbf{x}, \mathbf{p})}. \tag{4.7}$$

The last line denotes the expectation value according to the underlying distribution $W(\mathbf{x}, \mathbf{p})$.

Given an efficient way to approximately evaluate expectation values of quantum operators in phase space, we are interested in simulating dynamics in the phase-space representation. Therefore, we derive the equations of motion in the quantum phase-space language. Using the definition in Eq. (4.3) it can be shown that the Weyl symbol of an operator product is given by

$$(O_1 O_2)_{\mathrm{W}}(\mathbf{x}, \mathbf{p}) = O_{1,\mathrm{W}}(\mathbf{x}, \mathbf{p}) \exp\left[-\frac{i\hbar}{2}\Lambda\right] O_{2,\mathrm{W}}(\mathbf{x}, \mathbf{p}), \tag{4.8}$$

with the simplectic operator,

$$\Lambda = \frac{\overleftarrow{\partial}}{\partial\mathbf{p}}\frac{\overrightarrow{\partial}}{\partial\mathbf{x}} - \frac{\overleftarrow{\partial}}{\partial\mathbf{x}}\frac{\overrightarrow{\partial}}{\partial\mathbf{p}}, \tag{4.9}$$

where the arrows denote in which direction the derivatives are applied. This simplectic operator also appears in the definition of the classical Poisson bracket,

$$\{A, B\}_{\mathrm{PB}} = - A \Lambda B. \tag{4.10}$$

Given these expressions, we can derive the Weyl symbol of the commutator,

$$\hat{O}_{\mathrm{c}} := \left[\hat{O}_1, \hat{O}_2\right], \tag{4.11}$$

$$O_{\mathrm{c,W}} = (O_1 O_2)_{\mathrm{W}} - (O_2 O_1)_{\mathrm{W}}$$

$$= - 2i O_{1,\mathrm{W}} \sin\left[\frac{\hbar}{2}\Lambda\right] O_{2,\mathrm{W}} \tag{4.12}$$

$$= i\hbar \left\{O_{1,\mathrm{W}}, O_{2,\mathrm{W}}\right\}_{\mathrm{MB}}.$$

Here we leave out the dependencies for convenience sake and introduce the Moyal bracket,

$$\{A, B\}_{\mathrm{MB}} = - \frac{2}{\hbar} A \sin\left[\frac{\hbar}{2}\Lambda\right] B. \tag{4.13}$$

In the limit of small quantum fluctuations, $\hbar \to 0$, this Moyal bracket takes the form of the classical Poisson bracket,

$$\sin\left[\frac{\hbar}{2}\Lambda\right] = \frac{\hbar}{2}\Lambda + O\left(\hbar^3\right), \tag{4.14}$$

$$\{A, B\}_{\mathrm{MB}} = - A \Lambda B + O\left(\hbar^2\right)$$

$$= \{A, B\}_{\mathrm{PB}} + O\left(\hbar^2\right), \tag{4.15}$$

with big-O-notation $O(\hbar^2)$. The Weyl symbol of the commutator can thus be approximated by

$$O_{\mathrm{c,W}}(\mathbf{x}, \mathbf{p}) = i\hbar \left\{O_{1,\mathrm{W}}(\mathbf{x}, \mathbf{p}), O_{2,\mathrm{W}}(\mathbf{x}, \mathbf{p})\right\}_{\mathrm{PB}} + O\left(\hbar^3\right). \tag{4.16}$$

Given an expression for the Weyl symbol of the commutator, we can derive the representation of quantum dynamics in phase space. Let us therefore consider the von-Neumann equation,

$$i\hbar \frac{\partial \hat{\rho}(t)}{\partial t} = \left[\hat{H}(t), \hat{\rho}(t)\right], \tag{4.17}$$

which leads to the equation of motion for the Wigner function,

$$i\hbar\dot{W}\left(\mathbf{x}\left[t\right],\mathbf{p}\left[t\right]\right)=\left\{H_{\mathrm{W}}\left(\mathbf{x}\left[t\right],\mathbf{p}\left[t\right]\right),W\left(\mathbf{x}\left[t\right],\mathbf{p}\left[t\right]\right)\right\}_{\mathrm{MB}}$$
$$=\left\{H_{\mathrm{W}}\left(\mathbf{x}\left[t\right],\mathbf{p}\left[t\right]\right),W\left(\mathbf{x}\left[t\right],\mathbf{p}\left[t\right]\right)\right\}_{\mathrm{PB}}+O\left(\hbar^{3}\right). \tag{4.18}$$

As the commutator reduces to the classical Poisson bracket in the limit of small fluctuations, the Wigner function acts as a classical probability distribution and is conserved in time. Hence, classical trajectories of the quantum phase-space variables are given by the characteristics of Eq. (4.18), along which the Wigner function is conserved.

The expectation value of a quantum operator at a time $t$ can thus be approximated in a semi-classical manner,

$$\left\langle\hat{O}\left(\hat{\mathbf{x}},\hat{\mathbf{p}},t\right)\right\rangle\approx\int\int\mathrm{d}^{D}\mathbf{x}_{0}\mathrm{d}^{D}\mathbf{p}_{0}W\left(\mathbf{x}_{0},\mathbf{p}_{0}\right)O_{\mathrm{W}}\left(\mathbf{x}_{\mathrm{cl}}\left[t\right],\mathbf{p}_{\mathrm{cl}}\left[t\right]\right). \tag{4.19}$$

The indices "cl" denote classical time evolution and the index "0" denotes the initial values at $t = 0$. This way, a semi-classical approximation of quantum dynamics is given by sampling initial configurations from the Wigner function of the initial quantum system and evolving these samples classically in time. Averaging the outcomes at a desired time over the fluctuations in the initial configurations then provides the simulation outcome. This is the so-called truncated Wigner approximation, where the term "truncated" refers to the truncation of the equations of motion at first order in quantum fluctuations. This procedure gives insights into the relation between a classical and a quantum description and can in principle even be expanded to higher orders in $\hbar$, where the calculations get more complicated. Therefore, since higher order corrections can be calculated, the truncated Wigner approximation is controlled.

Nevertheless, the Wigner function is only a quasi-probability distribution, so the ansatz is still limited. As already discussed, the Wigner function is positive for most important classes of quantum systems, but if it takes negative values we cannot sample the initial states. Additionally, as the Wigner function is conserved during the classical time evolution, it can also not reach negative values at later times so that not all quantum states can be represented. This might lead to wrong correlation functions at later times in the simulations. The ansatz is used frequently for simulations of several quantum systems, e.g., bosonic systems at low temperatures [9–11]. Furthermore, it gives accurate results in many cases, especially at short time scales and for weakly interacting systems.

From now on we omit the indication of operators by hats again.

## 4.2 Discrete Quantum Phase Space

To represent discrete spin systems, such as spin-1/2 systems, in a quantum phase space, also the phase space can be chosen in a discrete manner. For the specific example of a spin-1/2 system, which we consider in the following, a discrete quantum

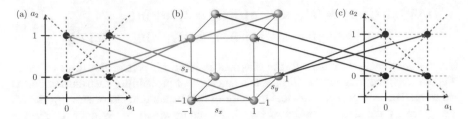

**Fig. 4.1** Setup of the discrete quantum phase space for a single spin-1/2 particle. Panels **a** and **c** show two different phase-space representations created by two different phase-point operators $A_\alpha$ from Eq. (4.21) and $A'_\alpha$ from Eq. (4.23). The dashed lines denote the three sets of two parallel lines each. The mapping between the spin state and the phase-space representation is pictured by the arrows pointing to the corners of the cube in panel (**b**), which denotes the eight possible configurations of the spin state with $s_\mu = \pm 1, \mu \in \{x, y, z\}$. Mapping the density matrix into phase space provides the Wigner function, a quasi-probability distribution underlying the phase-space points. In the case of a non-negative Wigner function spin configurations can be sampled from it. Figure adapted from [3]

phase space can be defined as a real-valued finite field spanned by four points. Each point represents one state in phase space and can be expressed as a two-dimensional vector $\boldsymbol{\alpha} = (a_1, a_2), a_1, a_2 \in \{0, 1\}$, over the finite field. The four points can then be connected via three sets of two parallel lines each, as illustrated in Fig. 4.1a, c [12, 13].

In the continuous case, one observable is associated with each set of parallel lines, where usually momentum **p** and position **x** are chosen for the horizontal and vertical lines, respectively. For the discrete case we analogously associate the Pauli operators $\sigma^x$, $\sigma^y$ and $\sigma^z$ with the three sets of parallel lines [12, 13]. Each line of each set is then identified with one eigenvalue of the corresponding operator. This is analogous to the continuous case and corresponds to the eigenvalues $\pm 1$ for the Pauli operators [12, 13]. The whole setup is illustrated in Fig. 4.1a, c.

To map operators $O$ from the Hilbert space onto the Weyl symbols $O_W$ in the discrete phase space, phase-point operators $A_\alpha$ can be defined [12, 13]. These map every point in the Hilbert space onto a point in the phase space, and the Weyl symbols are given by

$$O_W (\boldsymbol{\alpha}) = \frac{1}{2} \mathrm{Tr} \left[ O A_\alpha \right]. \tag{4.20}$$

Note that the choice of $A_\alpha$ is not unique [12–15]. A convenient choice for the phase-point operators is

$$A_\alpha = \frac{1}{2} \left[ (-1)^{a_1} \sigma^x + (-1)^{a_1 + a_2} \sigma^y + (-1)^{a_2} \sigma^z + \mathbb{1} \right], \tag{4.21}$$

which results in the mapping

$$\left(s_x = 1, s_y = 1, s_z = 1\right) \to \left(a_1 = 0, a_2 = 0\right),$$
$$\left(s_x = -1, s_y = -1, s_z = 1\right) \to \left(a_1 = 1, a_2 = 0\right),$$
$$\left(s_x = 1, s_y = -1, s_z = -1\right) \to \left(a_1 = 0, a_2 = 1\right),$$
$$\left(s_x = -1, s_y = 1, s_z = -1\right) \to \left(a_1 = 1, a_2 = 1\right),$$

$$(4.22)$$

where $s_\mu$ denote the eigenvalues of the Pauli operator $\sigma^\mu$, $\mu \in \{x, y, z\}$. The state of a single spin-1/2 particle can be uniquely described in terms of these eigenvalues, which yield $2^3 = 8$ possible states the spin can be found in. These can be depicted as corners of a cube, see Fig. 4.1b. The mapping defined via the phase-point operators $A_\alpha$ is depicted in Fig. 4.1a. This way, only four of the eight possible spin configurations shown in Fig. 4.1b are represented in the phase space [15].

Since the phase-point operators are not unique, additional operators can be derived via unitary transformations, $A'_\alpha = U A_\alpha U^\dagger$ [15]. One possible transformation leads to

$$A'_\alpha = \frac{1}{2}\left[(-1)^{a_1}\sigma^x - (-1)^{a_1+a_2}\sigma^y + (-1)^{a_2}\sigma^z + \mathbb{1}\right], \qquad (4.23)$$

giving the mapping

$$\left(s_x = 1, s_y = -1, s_z = 1\right) \to \left(a_1 = 0, a_2 = 0\right),$$
$$\left(s_x = -1, s_y = 1, s_z = 1\right) \to \left(a_1 = 1, a_2 = 0\right),$$
$$\left(s_x = 1, s_y = 1, s_z = -1\right) \to \left(a_1 = 0, a_2 = 1\right),$$
$$\left(s_x = -1, s_y = -1, s_z = -1\right) \to \left(a_1 = 1, a_2 = 1\right).$$

$$(4.24)$$

This maps the remaining four spin states onto a discrete phase-space representation, as depicted in Fig. 4.1c. By combining the two phase-space representations all spin configurations can be represented and with this the full single-particle spin state is expressed in terms of two discrete quantum phase spaces [15].

The Wigner function is then defined as the Weyl symbol of the density matrix,

$$W_\alpha = \frac{1}{2}\text{Tr}\left[\rho A_\alpha\right]. \qquad (4.25)$$

Similar to the continuous case, it shows properties of a probability distribution and defines a quasi-probability, which can take negative values, for each point in the phase space [2, 8, 12, 13]. If the Wigner function has only non-negative entries, it represents a proper probability distribution from which spin configurations can be sampled. This way, initial spin configurations can be drawn for each site of the spin system individually from a separate discrete phase space. However, since the discrete quantum phase spaces are defined for single spin particles, no (quantum) correlations between the spins can be represented. Thus, this ansatz can only be used to represent spin systems in separable states, where the full state can be expressed as a tensor product of the individual spin states.

For entangled states, it is in principle possible to construct a large discrete quantum phase space and sample full state configurations from it by mapping the total density matrix onto the phase space. However, since the density matrix scales exponentially with the system size, the size of the quantum phase space and with this the computational costs for this mapping also scales exponentially with the system size. This renders the ansatz inefficient and limits the quantum phase-space representation to separable states for large system sizes.

Combining the ansatz with a truncated time evolution of the initially sampled states provides the discrete truncated Wigner approximation (dTWA) analogously to the TWA [14–17].

## 4.3   Truncating the Time Evolution

Similar to the TWA, the initial configurations sampled from the discrete Wigner function in the dTWA can be evolved in time to simulate dynamics in spin systems. We consider the dynamics in the Heisenberg picture, so that the phase-point operators become time-dependent while the Wigner function remains stationary, and derive the truncated equations of motion within this section.

Since one discrete quantum phase space is defined for each site in the spin system, the total phase-point operator can be written as a tensor product of the individual phase-point operators,

$$A_\alpha = A_{\alpha_1} \otimes \cdots \otimes A_{\alpha_N}, \tag{4.26}$$

for a system with $N$ sites. The time-dependent phase-point operator $\mathcal{A}_{1,\dots,N}(t)$ is then given by applying unitary time-evolution operators solving the Schrödinger equation, as stated in Eq. (2.41),

$$\mathcal{A}_{1,\dots,N}(t) = \exp\left[-iHt\right]\left(A_{\alpha_1} \otimes \cdots \otimes A_{\alpha_N}\right)\exp\left[iHt\right]. \tag{4.27}$$

For convenience sake we drop the explicit notation of the time-dependence in the following equations. The equations of motion for the phase-point operators are given by means of the von-Neumann equation,

$$i\dot{\mathcal{A}}_{1,\dots,N} = \left[H, \mathcal{A}_{1,\dots,N}\right]. \tag{4.28}$$

This differential equation cannot be solved exactly in general and we thus need to approximate the time evolution. We do this by expanding $\mathcal{A}_{1,\dots,N}$ in a BBGKY-hierarchy (named after Bogoliubov, Born, Green, Kirkwood and Yvon) and truncating it at first or second order [15].

For the derivation of the truncated time evolution we introduce reduced phase-point operators analogously to reduced density matrices,

$$\mathcal{A}_{1,\ldots,s} = \mathrm{Tr}_{s+1,\ldots,N}\left[\mathcal{A}_{1,\ldots,N}\right]. \tag{4.29}$$

Furthermore, we define first-, second- and third-order reduced phase-point operators,

$$\mathcal{A}_j = \mathrm{Tr}_{\not{j}}\left[\mathcal{A}_{1,\ldots,N}\right], \tag{4.30}$$

$$\mathcal{A}_{j,k} = \mathrm{Tr}_{\not{j},\not{k}}\left[\mathcal{A}_{1,\ldots,N}\right], \tag{4.31}$$

$$\mathcal{A}_{j,k,l} = \mathrm{Tr}_{\not{j},\not{k},\not{l}}\left[\mathcal{A}_{1,\ldots,N}\right], \tag{4.32}$$

where the trace runs over all sites except for $j$, $j$ and $k$, or $j$, $k$ and $l$, respectively [15, 18, 19]. Considering a general Hamiltonian with on-site interaction part $H_j$ and pair-interaction part $H_{j,k}$,

$$H - \sum_{j=1}^{N} H_j + \sum_{j,k=1}^{N} H_{j,k}, \tag{4.33}$$

the equations of motion can be written down for $\mathcal{A}_j$ and $\mathcal{A}_{j,k}$ [15],

$$i\dot{\mathcal{A}}_j = \left[H_j, \mathcal{A}_j\right] + \sum_{\substack{l=1 \\ l \neq j}}^{N} \mathrm{Tr}_l\left(\left[H_{j,l}, \mathcal{A}_{j,l}\right]\right), \tag{4.34}$$

$$i\dot{\mathcal{A}}_{j,k} = \left[H_j + H_k + H_{j,k}, \mathcal{A}_{j,k}\right] + \sum_{\substack{l=1 \\ l \neq j,k}}^{N} \mathrm{Tr}_l\left(\left[H_{j,l} + H_{k,l}, \mathcal{A}_{j,k,l}\right]\right). \tag{4.35}$$

These expressions can be derived from Eq. (4.28). To get expressions only in terms of first-order reduced phase-point operators, we introduce correlation operators $C_{j,k}$, $C_{j,k,l}$ such that

$$\mathcal{A}_{j,k} = \mathcal{A}_j\mathcal{A}_k + C_{j,k}, \tag{4.36}$$

$$\mathcal{A}_{j,k,l} = \mathcal{A}_j\mathcal{A}_k\mathcal{A}_l + \mathcal{A}_j C_{k,l} + \mathcal{A}_k C_{j,l} + \mathcal{A}_l C_{j,k} + C_{j,k,l}. \tag{4.37}$$

Plugging these expressions into Eqs. (4.34)–(4.35), we get

$$i\dot{\mathcal{A}}_j = \left[H_j, \mathcal{A}_j\right] + \sum_{\substack{l=1 \\ l \neq j}}^{N} \mathrm{Tr}\left(\left[H_{j,l}, \mathcal{A}_j\mathcal{A}_l\right] + \left[H_{j,l}, C_{j,l}\right]\right), \qquad (4.38)$$

$$i\dot{C}_{j,k} = \left[H_j + H_k, C_{j,k}\right] + \left[H_{j,k}, \mathcal{A}_j\mathcal{A}_k + C_{j,k}\right]$$

$$\sum_{\substack{l=1 \\ l \neq j,k}}^{N} \mathrm{Tr}_l\left(\left[H_{j,l}, \mathcal{A}_jC_{k,l}\right]\right) + \sum_{\substack{l=1 \\ l \neq j,k}}^{N} \mathrm{Tr}_l\left(\left[H_{k,l}, \mathcal{A}_kC_{j,l}\right]\right)$$

$$\sum_{\substack{l=1 \\ l \neq j,k}}^{N} \mathrm{Tr}_l\left(\left[H_{j,l} + H_{k,l}, \mathcal{A}_lC_{j,k}\right]\right) - \mathcal{A}_k\mathrm{Tr}_k\left(\left[H_{j,k}, \mathcal{A}_j\mathcal{A}_k + C_{j,k}\right]\right) \quad (4.39)$$

$$- \mathcal{A}_j\mathrm{Tr}_j\left(\left[H_{k,j}, \mathcal{A}_k\mathcal{A}_j + C_{k,j}\right]\right) + \sum_{\substack{l=1 \\ l \neq j,k}}^{N} \left(\mathrm{Tr}_l\left[H_{j,l}, C_{j,k,l}\right]\right.$$

$$+ \mathrm{Tr}_l\left[H_{k,l}, C_{j,k,l}\right]\right).$$

Truncating the BBGKY-hierarchy at first order sets $C_{j,k} = 0$ for all $j, k$ and gives the equations of motion

$$i\dot{\mathcal{A}}_j = \left[H_j, \mathcal{A}_j\right] + \sum_{\substack{l=1 \\ l \neq j}}^{N} \mathrm{Tr}_l\left(\left[H_{j,l}, \mathcal{A}_j\mathcal{A}_l\right]\right), \qquad (4.40)$$

which are the mean-field equations of motion [2]. A truncation at second order is achieved by setting $C_{j,k,l} = 0$ for all $j, k, l$ and can be read from Eqs. (4.38)–(4.39) [15].

If we consider a spin-1/2 system, as we do in the following, the phase-point operators and correlation operators can be expressed in terms of the Pauli matrices $\sigma_j^\mu$,

$$\mathcal{A}_j = \frac{1}{2}\left(\mathbb{1} + \mathbf{a}_j\boldsymbol{\sigma}_j\right), \qquad (4.41)$$

$$C_{j,k} = \frac{1}{4}\sum_{\mu,\nu\in\{x,y,z\}} c_{j,k}^{\mu,\nu}\sigma_j^\mu\sigma_k^\nu, \qquad (4.42)$$

with expansion coefficients $\mathbf{a}_j$ and $c_{j,k}$ [15]. A general Hamiltonian with on-site and pair interactions for such a system can also be written in terms of the Pauli matrices,

$$H = \sum_{j=1}^{N} H_j + \sum_{j,k=1}^{N} H_{j,k}$$

$$= \sum_{j=1}^{N} \mathbf{h}_j\boldsymbol{\sigma}_j + \sum_{j,k=1}^{N}\sum_{\beta\in\{x,y,z\}} J_{j,k}^\beta\sigma_j^\beta\sigma_k^\beta. \qquad (4.43)$$

Plugging Eqs. (4.41)–(4.43) into Eqs. (4.38)–(4.39), the first-order equations of motion for the expansion coefficients result in

$$\dot{a}_j^\mu = -2 \sum_{\gamma,\delta\in\{x,y,z\}} \left[ h_j^\gamma a_j^\delta \epsilon^{\mu\gamma\delta} + \sum_{\substack{l=1\\l\neq j}}^N J_{j,l}^\gamma a_j^\delta a_l^\gamma \epsilon^{\mu\gamma\delta} \right], \qquad (4.44)$$

which are the classical equations of motion for a spin system [14–16]. The second-order equations of motion then yield [15]

$$\dot{a}_j^\mu = -2 \sum_{\gamma,\delta\in\{x,y,z\}} \left[ h_j^\gamma a_j^\delta \epsilon^{\mu\gamma\delta} + \sum_{\substack{l=1\\l\neq j}}^N J_{j,l}^\gamma \left( a_j^\delta a_l^\gamma \epsilon^{\mu\gamma\delta} + c_{j,l}^{\delta,\gamma} \epsilon^{\mu\gamma\delta} \right) \right], \qquad (4.45)$$

$$
\begin{aligned}
\dot{c}_{j,k}^{\mu,\nu} = 2 \Bigg\{ &- \sum_{\gamma\in\{x,y,z\}} \left( J_{j,k}^\nu a_j^\gamma - J_{j,k}^\mu a_k^\gamma \right) \epsilon^{\mu\nu\gamma} \\
&- \sum_{\delta,\gamma\in\{x,y,z\}} \left[ \left( h_j^\gamma + \sum_{\substack{l=1\\l\neq j,k}}^N J_{j,l}^\gamma a_l^\gamma \right) c_{j,k}^{\delta,\nu} \epsilon^{\gamma\delta\mu} + \left( h_k^\gamma + \sum_{\substack{l=1\\l\neq j,k}}^N J_{k,l}^\gamma a_l^\gamma \right) c_{j,k}^{\mu,\delta} \epsilon^{\gamma\delta\nu} \right] \\
&- \sum_{\delta,\gamma\in\{x,y,z\}} \sum_{\substack{l=1\\l\neq j,k}}^N \left[ J_{j,l}^\gamma a_j^\delta c_{k,l}^{\nu\gamma} \epsilon^{\gamma\delta\mu} + J_{k,l}^\gamma a_k^\delta c_{j,l}^{\mu\gamma} \epsilon^{\gamma\delta\nu} \right] \\
&\sum_{\delta,\gamma\in\{x,y,z\}} J_{j,k}^\gamma \left[ a_j^\mu \left( c_{j,k}^{\gamma,\delta} + a_j^\gamma a_k^\delta \right) \epsilon^{\gamma\delta\nu} + a_k^\nu \left( c_{j,k}^{\delta\gamma} + a_j^\delta a_k^\gamma \right) \epsilon^{\gamma\delta\mu} \right] \Bigg\}.
\end{aligned}
$$

$$(4.46)$$

Expectation values of one- and two-spin functions can be calculated from the expansion coefficients as

$$\langle \sigma_j^\mu \rangle \approx \frac{1}{Q} \sum_{q=1}^Q a_{j;q}^\mu, \qquad (4.47)$$

$$\langle \sigma_j^\mu \sigma_k^\nu \rangle \approx \frac{1}{Q} \sum_{q=1}^Q \left( c_{j,k;q}^{\mu,\nu} + a_{j;q}^\mu a_{k;q}^\nu \right). \qquad (4.48)$$

$Q$ is the total number of initial configurations sampled from the Wigner function and the index $q$ denotes the expansion coefficient resulting from sample $q$ [15]. For first-order truncated simulations, expectation values of two-spin functions can be evaluated via

$$\langle \sigma_j^\mu \sigma_k^\nu \rangle \approx \frac{1}{Q} \sum_{q=1}^{Q} \mathfrak{a}_{j;q}^\mu \mathfrak{a}_{k;q}^\nu, \tag{4.49}$$

as the second-order expansion coefficients are set to zero.

Thus, in the dTWA a set of initial configurations of expansion coefficients $\mathfrak{a}_j^\mu$ is sampled from the Wigner function on the discrete phase space and each configuration is evolved in time, where the time evolution is truncated at first or second order. The expansion coefficients $c_{j,k}^{\mu,\nu}$ are set to zero as initial configuration. By averaging the resulting observables at different times, dynamics of expectation values can be simulated approximately in a semi-classical manner [14–17].

## 4.4 Simulating Sudden Quenches in the TFIM

### 4.4.1 Preparing the Ground State

Having introduced the rudiments of the dTWA, we are interested in benchmarking it on sudden quenches in the transverse-field Ising model (TFIM). Thus, we take a spin-1/2 system and prepare it in the ground state of a TFIM Hamiltonian with initial transverse field $h_{t,i}$,

$$H_{\text{TFIM}} = - \sum_{i=1}^{N} \sigma_i^x \sigma_{(i+1)\bmod N}^x - h_{t,i} \sum_{i=1}^{N} \sigma_i^z, \tag{4.50}$$

with $(i+1)\bmod N$ denoting modulo $N$ calculations and implying periodic boundary conditions. $N$ denotes the chain length. We then abruptly bring the system out of equilibrium by changing the transverse field to a final value $h_{t,f}$, describing a sudden quench. The dynamics in the system after the quench can be simulated using the dTWA and the results can be compared with exact solutions as discussed in Sect. 2.5.2 [20–23].

As an initial state we choose the ground state of a TFIM Hamiltonian with infinitely large transverse field, $h_{t,i} \to \infty$, which is the fully $z$-polarized state,

$$|\Psi\rangle = \bigotimes_{i=1}^{N} |\uparrow\rangle_i, \tag{4.51}$$

with $|\uparrow\rangle_i$ denoting the spin at site $i$ in the up-state. For this state the single-spin density matrix reads

$$\rho = |\uparrow\rangle\langle\uparrow|$$

$$= \begin{pmatrix} 1 \\ 0 \end{pmatrix} (1 \; 0) \tag{4.52}$$

$$= \begin{bmatrix} 1 & 0 \\ 0 & 0 \end{bmatrix}.$$

We can now calculate the Wigner functions on the two discrete phase spaces introduced in Sect. 4.2 via Eq. (4.25) using the corresponding phase-point operators [12–17]. For the first discrete phase space, defined by the phase-point operators $\mathcal{A}_\alpha$ [Eq. (4.21)], we get

$$
\begin{aligned}
W_{(0,0)} &= \frac{1}{2} &\leftrightarrow& \quad (s_x = 1, s_y = 1, s_z = 1), \\
W_{(1,0)} &= \frac{1}{2} &\leftrightarrow& \quad (s_x = -1, s_y = -1, s_z = 1), \\
W_{(0,1)} &= 0 &\leftrightarrow& \quad (s_x = 1, s_y = -1, s_z = -1), \\
W_{(1,1)} &= 0 &\leftrightarrow& \quad (s_x = -1, s_y = 1, s_z = -1).
\end{aligned}
\tag{4.53}
$$

For convenience sake we state the spin state corresponding to the considered point in phase space. For the second phase space, defined by the phase-point operators $\mathcal{A}'_\alpha$ [Eq. (4.23)], we get the Wigner function [15]

$$
\begin{aligned}
W'_{(0,0)} &= \frac{1}{2} &\leftrightarrow& \quad (s_x = 1, s_y = -1, s_z = 1), \\
W'_{(1,0)} &= \frac{1}{2} &\leftrightarrow& \quad (s_x = -1, s_y = 1, s_z = 1), \\
W'_{(0,1)} &= 0 &\leftrightarrow& \quad (s_x = 1, s_y = 1, s_z = -1), \\
W'_{(1,1)} &= 0 &\leftrightarrow& \quad (s_x = -1, s_y = -1, s_z = -1).
\end{aligned}
\tag{4.54}
$$

This shows that the initial state for the dTWA is sampled equally from the four states

$$
\left\{ \begin{pmatrix} s_x = 1 \\ s_y = 1 \\ s_z = 1 \end{pmatrix}, \begin{pmatrix} s_x = -1 \\ s_y = -1 \\ s_z = 1 \end{pmatrix}, \begin{pmatrix} s_x = 1 \\ s_y = -1 \\ s_z = 1 \end{pmatrix}, \begin{pmatrix} s_x = -1 \\ s_y = 1 \\ s_z = 1 \end{pmatrix} \right\}, \tag{4.55}
$$

which are all states with $s_z = 1$. For this initial state the Wigner function is always positive, so that it is a probability distribution and we can sample the initial spin configurations. We can then evolve these samples in time by numerically integrating Eq. (4.44) or Eqs. (4.45)–(4.46), providing simulations truncated at first or second order, respectively. For this we plug the Hamiltonian, Eq. (4.50), with the final transverse field $h_{t,f}$ into the equations of motion and use a fourth order Runge–Kutta scheme for the integration [24].

### 4.4.2  Dynamics of Correlations

As an observable for benchmarking the dTWA simulation of dynamics after sudden quenches in the TFIM we choose the correlation function in the $x$-basis,

$$C_d^{xx}\left(t, h_{\mathrm{t,f}}\right) = \langle \sigma_0^x\left(t\right) \sigma_d^x\left(t\right)\rangle, \tag{4.56}$$

where $d$ is the distance between the spins whose correlation is calculated. The correlation function is evaluated at time $t$ after a sudden quench to the final transverse field $h_{\mathrm{t,f}}$ from a fully $z$-polarized initial state. We can compare the dTWA simulations with exact solutions as discussed in Sect. 2.5.2, where the initial state is an effectively fully $z$-polarized state with $h_{\mathrm{t,i}} = 100$. We have checked that this state approximates the initial state chosen in the dTWA simulations with good accuracy.

Due to translation invariance in the spin chain with periodic boundary conditions, we can without loss of generality choose one of the two considered spins at position $i = 0$ when calculating the exact correlations. The outcome only depends on the distance between the two spins and not on their absolute positions. In the numerical simulations, translation invariance is not exactly given. We can thus average over all positions to get better statistics,

$$C_{d,\mathrm{dTWA}}^{xx}\left(t, h_{\mathrm{t,f}}\right) = \frac{1}{N} \sum_{i=1}^{N} \langle \sigma_i^x\left(t\right) \sigma_{(i+d)\bmod N}^x\left(t\right)\rangle. \tag{4.57}$$

In the following we compare the exact dynamics after sudden quenches within the paramagnetic phase ($h_{\mathrm{t}} > 1$) with dTWA results, truncating at first and second order using two different sample sets for the initial states. In the first sampling scheme we consider the initial configurations $S_4$, which are obtained by applying only the phase-point operator $A_\alpha$ [Eq. (4.21)], as introduced in Sect. 4.2. These are only four configurations, so that only half of the state space is covered and only two states with $s_z = 1$ are drawn. In the second sampling scheme, we use the initial configurations $S_8$, which are the spin states obtained when applying the two sets of phase-point operators $A_\alpha$ and $A_\alpha'$ [Eq. (4.23)], as depicted in Fig. 4.1. Here the whole state space is covered.

Figure 4.2 shows the absolute value of the correlation function in a spin chain with $N = 20$ sites at three different times after a sudden quench to $h_{\mathrm{t,f}} = 1.0001$, very close to the quantum critical point. The same-site correlation function at $d = 0$ is given by the normalization of the state due to $\sigma_i^x \sigma_i^x = \mathbb{1}$ [15]. The initial state is reproduced exactly up to statistical fluctuations due to the finite sample size, as it is sampled from the exact underlying distribution. Here we choose $R = 10000$ samples for the first-order and $R = 1000$ samples for the second-order calculations. At later times, deviations from the exact solution appear, which result from using the semi-classical approximation of the dynamics. In the calculations with truncation at second order, these deviations are smaller than in the results with truncation at

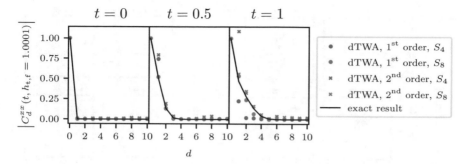

**Fig. 4.2** Correlation function $C_d^{xx}(t, h_{t,f})$ [Eq. (4.57)] at three times $t = 0$ [(**a**)], $t = 0.5$ [(**b**)] and $t = 1$ [(**c**)] after a sudden quench to $h_{t,f} = 1.0001$. Discrete truncated Wigner approximation (dTWA) simulations of first (dots) and second (crosses) order with the initial state sampled from $S_4$ (blue, single phase space) and $S_8$ (orange, two combined phase spaces) are plotted together with the exact solution (black line) as a function of the relative distance $d$ between the spins. Error bars are of the size of data points and hence not shown. The exact solution is captured well at short times, especially when sampling from $S_8$. Figure adapted from [3]

first order. This comes expected since the equations of motion are closer to the exact quantum ones.

One can already see in Fig. 4.2 that the correlation function decays exponentially with distance $d$ at later times after the quench. For a better visualization, Fig. 4.3 shows the absolute value of the correlation function in a log-scale [Fig. 4.3a]. Moreover, it shows the absolute error [Fig. 4.3b],

$$\Delta_d^{xx}(t, h_{t,f}) = \left| C_{d,\text{exact}}^{xx}(t, h_{t,f}) - C_{d,\text{dTWA}}^{xx}(t, h_{t,f}) \right|, \tag{4.58}$$

and the residual [Fig. 4.3c],

$$R_d^{xx}(t, h_{t,f}) = \left| \frac{C_{d,\text{exact}}^{xx}(t, h_{t,f}) - C_{d,\text{dTWA}}^{xx}(t, h_{t,f})}{C_{d,\text{exact}}^{xx}(t, h_{t,f})} \right|, \tag{4.59}$$

for different times $t$ after a sudden quench to $h_{t,f} = 1.0001$. In Fig. 4.3a one can observe that the dTWA results reach a plateau at correlations of a size $C_{d,\text{dTWA}}^{xx} \approx 10^{-3}$, where the exact solution at short times is much smaller. This plateau is due to statistical fluctuations and results from the finite sample size, so that no smaller values can be reached. Thus, these data points are rendered transparent in Fig. 4.3, as they cannot represent the exact solution.

The plateau can be shifted to smaller values by increasing the sample size $R$. The analysis of the scaling of the plateau with $R$ shows that it goes as $\propto R^{-1/2}$, what is expected from general statistical arguments [25]. To capture correlations of the size $C_d^{xx} \approx 10^{-5}$, we would need $R \approx 10^6$ samples, which leads to excessively large computation times. Thus, we cannot increase $R$ far enough to capture the exponential decay at large distances, but we have to focus on the short-distance behavior. It is

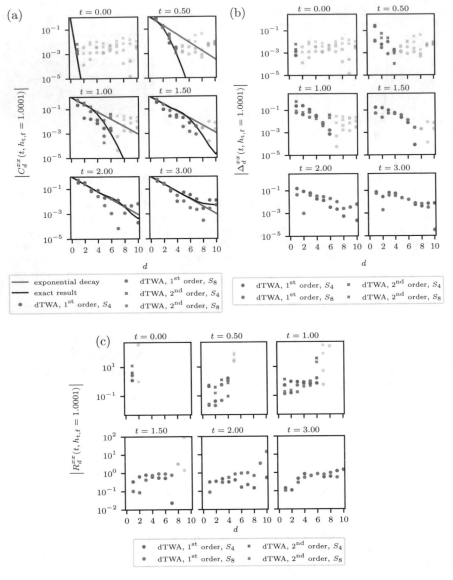

**Fig. 4.3** Correlation function Eq. (4.57) [(**a**)], absolute deviation Eq. (4.58) [(**b**)] and relative error Eq. (4.59) [(**c**)] at different times after a sudden quench to $h_{t,f} = 1.0001$. DTWA simulations with initial samples from $S_4$ (blue) and $S_8$ (orange), and truncation at first (dots) and second (crosses) order are plotted as a function of relative distance $d$. In panel **a** the exact solution is plotted for comparison (black line) together with an exponential function fitted to the first three points (green line). Points appearing in the limiting plateau due to finite sample sizes are transparent. Error bars are of the size of data points. Figure adapted from [3]

hence sufficient to consider spin chains with $N = 20$ sites. If not stated otherwise, all following simulations within this chapter are done for $N = 20$ sites and $R = 10000$ samples for the first-order or $R = 1000$ samples for the second-order dTWA.

At times $t \approx 1$, one finds clear deviations from the exact solution for the first order simulations in Fig. 4.3, while the second-order simulations still describe the dynamics well, as can be observed more clearly in Fig. 4.3b and c. However, for later times the second-order results are not shown anymore. They have diverged due to instabilities in the equations of motion, as discussed in [26, 27]. For this reason, we focus on the first-order simulations in the following, as we are interested in late-time dynamics.

In Fig. 4.3a an exponential function is fitted to the first three points of the exact correlation function and plotted as a green line. One can see that at late times the correlation function decays exponentially with distance $d$, since it converges to the exponential fit even for larger $d$. The deviations at large distances for $t \geq 3$ result from finite-size effects appearing at these times already.

The absolute errors in Fig. 4.3b are always around $\lesssim 0.1$, while the relative error in Fig. 4.3c diverges for large $d$, so that it is not plotted anymore, especially for short times after the quench. These divergences result from the plateau due to finite sampling. Even at later times it can be observed that the relative error for the first two points is smaller than for larger $d$.

We now investigate how the performance of the dTWA depends on the final transverse field $h_{t,f}$ after the quench. Therefore, Fig. 4.4 shows the correlation function [Fig. 4.4a], as well as the absolute deviation from the exact solution [Fig. 4.4b] and the relative errors [Fig. 4.4c] at a fixed time $t = 2$ after sudden quenches to different $h_{t,f}$. At these times, the second-order simulations have already diverged due to instabilities and are hence not shown anymore. One finds that the absolute error is again around $\lesssim 0.1$ for all $h_{t,f}$, and even the relative errors show the same behavior as discussed earlier. The residuals of the first two points are small, whereas for larger distances $d$ the relative deviation grows.

Again, an exponential function fitted to the first three points of the exact correlation function is shown in green in Fig. 4.4a. One can see that for short $d$ this decay is always represented by the correlation function. For quenches to larger values of $h_{t,f}$, a plateau, which does not follow the exponential function anymore, builds up at longer relative distances $d$ in the correlations. Here one finds the second exponential behavior at intermediate distances, as discussed in Sect. 2.5.2. The dTWA captures the plateau at large relative distances $d$ in the exact solution for larger values of $h_{t,f}$ well, while the relative errors are still at the order of $\sim 1$. This we expect to be improvable by increasing the sample size $R$. Given our computational resources, this is out of reach for now, as discussed earlier.

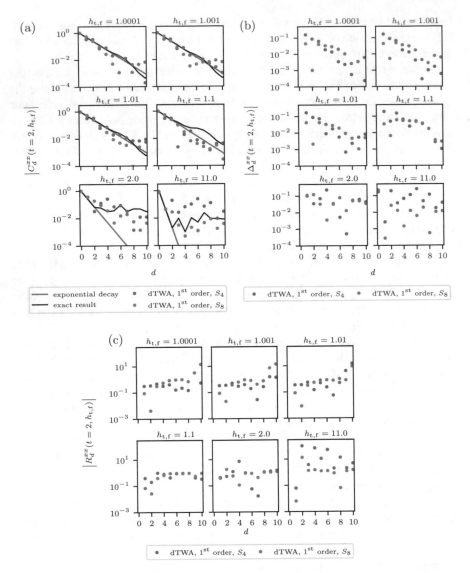

**Fig. 4.4** Correlation function [**a**, Eq. (4.57)], absolute deviation [**b**, Eq. (4.58)] and relative error [**c**, Eq. (4.59)] at a fixed time $t = 2$ after sudden quenches to different transverse fields $h_{t,f}$ within the paramagnetic regime. The first order dTWA with initial sampling from $S_4$ (blue) and $S_8$ (orange) is shown as a function of relative distance $d$, while second-order dTWA simulations have diverged at these times. In **a** the simulations are plotted on a semi-logarithmic scale together with the exact solution (black line) and an exponential function (green line) fitted to points at $d < 3$. Error bars are of the size of data points and hence not shown. Figure adapted from [3]

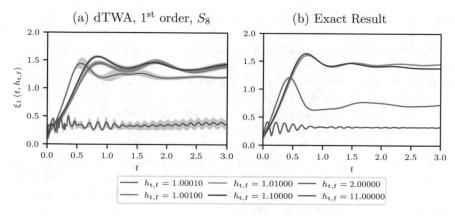

**Fig. 4.5** Correlation length $\xi_1(t, h_{t,f})$ [Eq. (4.60)] as a function of time after sudden quenches to different final transverse fields $h_{t,f}$ within the paramagnetic phase (different colors). Panel **b** shows dTWA simulations of first order with initial sampling from $S_8$, while panel **b** shows the exact solution. For quenches to intermediate distances from the quantum critical point large deviations are found in the long-time limit. Shaded regions denote the fitting uncertainty when extracting the correlation length. Figure adapted from [3]

### 4.4.3 Correlation Length: Short-Distance Behavior

From the exponential decay in the correlation function at short distances we can extract the correlation length $\xi_1$, as introduced in Sect. 2.5.2 [21, 22, 28, 29],

$$C_d^{xx}\left(t, h_{t,f}\right) \propto \exp\left[-\frac{d}{\xi_1\left(t, h_{t,f}\right)}\right], \quad \text{for } d < 3. \qquad (4.60)$$

Here we focus on the behavior of only the first three points, since in this regime the exponential decay is still present for quenches to larger final transverse fields. By considering the correlation length $\xi_1$ we can analyze the accuracy of the dTWA at short relative distances. To obtain the dynamics of the correlation length we fit an exponential function to the decay in the correlation function at distances $d < 3$ for different times after sudden quenches to different transverse fields $h_{t,f}$ [21, 22, 28, 29].

Figure 4.5 shows the time evolution after quenches within the paramagnetic regime to different distances from the quantum critical point, where the same values as in Fig. 4.4 are chosen for the transverse field. Figure 4.5a shows the dTWA simulations and Fig. 4.5b shows the exact result, where the correlation length is extracted from a fit to the short-distance behavior of the exact correlation function. The shaded regions in Fig. 4.5a denote the fitting uncertainty.

The exact short-time behavior is captured well in the dTWA results. This comes expected, since we have already observed in Fig. 4.4 that the simulations represent the exact correlation function very well at short times and small relative distances between the spins. The correlation length only reflects this short-distance behavior.

Around the first maximum of the damped oscillations observed in the correlation length dynamics deviations from the exact solution start to appear in the simulations. Even though the dTWA does not capture these oscillations well, it saturates at approximately the same values as the exact solution at late times, especially for quenches close to the quantum critical point. At intermediate distances, here for $h_{t,f} = 2$, the long-time behavior deviates strongly from the exact solution, while it gets better again even further away from the quantum critical point.

As discussed in Sect. 2.5.2, for the case of taking the initial transverse field in the limit of infinity, the late-time correlation length can be described by a generalized Gibbs ensemble (GGE). At stationarity, the correlation length is thus expected to follow the universal function [21, 29]

$$\xi_{GGE}(\epsilon) = \frac{1}{\ln(2\epsilon + 2)}, \tag{4.61}$$

with distance $\epsilon$ of the final transverse field from the quantum critical point,

$$\epsilon = \frac{h_{t,f} - h_{t,c}}{h_{t,c}} \tag{4.62}$$
$$= h_{t,f} - 1.$$

To check if the dTWA can capture the convergence to $\xi_{GGE}(\epsilon)$ at late times, Fig. 4.6 shows the exact and simulated correlation lengths at a fixed time $t = 2$ after sudden quenches as a function of the final distance $\epsilon$ from the quantum critical point on a logarithmic scale, together with $\xi_{GGE}(\epsilon)$. The correlation length saturates at a finite value at the quantum critical point and does not diverge, which is due to the fact that the quantum phase transition vanishes for non-zero temperatures in the one-dimensional TFIM [29].

One can observe that for small and large $\epsilon$ both the exact solution and the simulations have converged to $\xi_{GGE}(\epsilon)$. At intermediate distances stationarity is not yet reached, not even in the exact solution. To show this more clearly, we also plot the exact solution at a later time $t = 18$ after the quench, where we have to choose a larger chain size to avoid finite-size effects. It can be observed that the exact solution converges to $\xi_{GGE}(\epsilon)$ for all quenches at long enough times, as discussed in Sect. 2.5.2 [28, 29]. However, we find deviations in the dTWA from the exact result even at later times, so that these do not converge to $\xi_{GGE}(\epsilon)$ at intermediate distances from the quantum critical point.

To also consider quenches across the quantum critical point, Fig. 4.7 shows the correlation length at a fixed time $t = 2$ after a sudden quench to different final transverse fields $h_{t,f}$. Quenches are considered within the paramagnetic and into the ferromagnetic phase. The shaded regions denote the temporal oscillations around the final value appearing at earlier times, while the dashed vertical line indicates the quantum phase transition. Small transverse fields can be considered as perturbations in

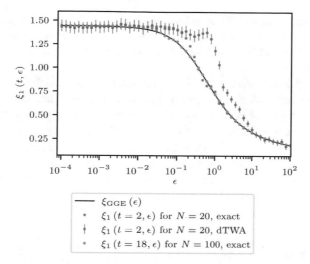

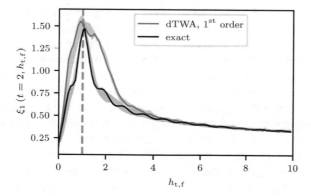

**Fig. 4.6** Correlation length $\xi_1(t, \epsilon)$ [Eq. (4.60)] at a fixed time $t$ after sudden quenches within the paramagnetic regime as a function of the final distance $\epsilon$ from the quantum critical point [Eq. (4.62)]. DTWA simulations of first order with initial sampling from $S_8$ (blue dots) are compared to exact solutions (orange dots) for a spin chain with $N = 20$ sites at $t = 2$ on a semi-logarithmic scale. The behavior of the GGE, which is expected for the exact solution at late times, is also shown according to Eq. (4.61) (black line). This behavior is not captured in the dTWA simulations at intermediate $\epsilon$. For completeness the exact solution of a spin chain with $N = 100$ sites at time $t = 18$ after the quenches is plotted to show the convergence to the GGE-function at late times in large systems (green dots). Error bars denote the fitting uncertainty when extracting the correlation length. Figure adapted from [3]

**Fig. 4.7** Correlation length $\xi_1(t, h_{t,f})$ [Eq. (4.60)] at a fixed time $t = 2$ after sudden quenches within the paramagnetic and into the ferromagnetic regime. DTWA simulations (green line) are plotted together with the exact solution (black line). The dashed vertical line denotes the quantum phase transition. Shaded regions show the amplitude of temporal oscillations in the correlation lengths at times between $t = 1.3$ and $t = 2$, where upper and lower limits correspond to maximum and minimum values reached, respectively. The dTWA shows deviations from the exact solution in the quantum critical regime in both phases. Figure adapted from [3]

the classical Ising model without a transverse field, which explains why the dTWA captures the exact solution very well in this region. These dynamics can also be described via a low-order cumulant expansion, which corresponds to a perturbation approach [30].

One finds the deviations from the exact solution around the quantum critical point as discussed earlier in the paramagnetic regime, but they also appear in the ferromagnetic regime. Further away from the quantum critical point, for large $h_{t,f}$, the simulations represent the exact solution well again. From these observations we conclude that the dTWA is limited in the vicinity of the quantum critical point, where large correlation lengths and hence strong long-range correlations appear. Furthermore, quantum entanglement is expected in this regime, which is here not captured in the semi-classical simulations.

## 4.5  Summary

In this chapter we have pointed out the limitations of the semi-classical discrete truncated Wigner approximation (dTWA) in a quantum critical regime by benchmarking the approach on sudden quenches in the transverse-field Ising model (TFIM). We have studied dynamics after sudden quenches from an effectively fully polarized state into the vicinity of the quantum critical point. Hereby we have focused on the regime around the quantum critical point. Strong long-range interactions appear in this region and the half-chain entanglement entropy grows linearly with time up to a saturation value that scales extensively with the system size (volume-law), see Sect. 2.5.2. As this is a regime where other existing simulation methods, such as the time-dependent density-matrix renormalization group, struggle, it is even more important to find simulation methods valid in this quantum critical region.

Considering the dTWA with equations of motion truncated at first and second order, we have shown that at short times after the quench both approaches show a good agreement with the exact solution. The second-order simulations have turned out to be more accurate than the first-order ones. However, we found the equations of motion truncated at second order to experience instabilities at later times. These led to divergences in the simulated correlation functions.

We have shown that the first-order calculations provide stable results even at later times and that they reproduce the exact correlation function well for quenches both close to the quantum critical point and far away from it. Nevertheless, they show deviations at intermediate distances. We have found this behavior especially for short-distance correlations. The dTWA has turned out to show a restriction to a minimum reachable value of the correlation function. Due to statistical errors resulting from the finite sample size, the simulated correlations cannot reach values below this limit. The lower bound can be shifted to smaller values by increasing the sample size, but we have found that it is computationally very expensive to reach smaller correlations. This has limited our simulations to regimes with large correlations, which is the short-distance regime in the exponentially decaying correlation function.

As the correlation length $\xi_1$ extracted from the correlation function in the TFIM is characterized by the short-distance behavior, see Sect. 2.5.2, we have shown that it is also captured well by the first-order dTWA. However, we have not found the correlation length $\xi_2$ at large relative distances in the simulated results. This can be extracted from the secondary oscillatory behavior with exponential envelope for quenches further away from the quantum critical point, as discussed in Sect. 2.5.2 [21, 29]. The second correlation length probably goes beyond the capabilities of the semi-classical simulations. It might require more quantum effects to be included in the approximation method, such as the conservation of additional quantities [31]. This is not necessarily given in the simulations. Furthermore, this correlation length appears at larger relative distances in the correlation function, which we have pointed out to be hard to capture in the dTWA due to finite sample size.

We have shown that the short-time dynamics of the correlation length are captured well by the semi-classical simulations for all quenches. However, when considering the late-time saturation, we only found accurate results close to the quantum critical point and far away from it. For quenches to intermediate distances from the quantum phase transition, $0.1 \lesssim \epsilon = h_{t,f} - 1 \lesssim 10$, the simulations have saturated at different values than the exact solution. Because of that, the behavior of a generalized Gibbs ensemble found in the long time limit of the exact correlation length could not be captured by the first-order dTWA.

In summary, we have pointed out the limitations of the dTWA in the vicinity of the quantum critical point, in regimes of strong long-range correlations. At the same time, we have shown that it captures the exact dynamics well far enough away from the quantum phase transition even for quenches across the quantum critical point into the ferromagnetic regime. We have furthermore found the short-time dynamics to be captured well for all quenches.

# References

1. Blakie PB, Bradley AS, Davis MJ, Ballagh RJ, Gardiner CW (2008) Dynamics and statistical mechanics of ultra-cold bose gases using c-field techniques. Adv Phys 57(5):363–455. https://doi.org/10.1080/00018730802564254
2. Polkovnikov A (2010) Phase space representation of quantum dynamics. Ann Phys 325(8):1790. https://doi.org/10.1016/j.aop.2010.02.006
3. Czischek Stefanie, Gärttner Martin, Oberthaler Markus, Kastner Michael, Gasenzer Thomas (2018) Quenches near criticality of the quantum Ising chain-power and limitations of the discrete truncated Wigner approximation. Quantum Sci Technol 4(1):014006. http://stacks.iop.org/2058-9565/4/i=1/a=014006
4. Scully MO, Zubairy MS (1997) Quantum optics. Cambridge University Press, Cambridge. https://doi.org/10.1017/CBO9780511813993
5. O'Connell RF, Wigner EP (1981) Quantum-mechanical distribution functions: Conditions for uniqueness. Phys Lett A 83(4):145–148. http://www.sciencedirect.com/science/article/pii/0375960181908707
6. Hillery M, O'Connell RF, Scully MO, Wigner EP (1984) Distribution functions in physics: fundamentals. Phys Rep 106(3):121–167. http://www.sciencedirect.com/science/article/pii/0370157384901601

7. Weyl H (1927) Quantenmechanik und Gruppentheorie. Z Phys 46(1):1–46. https://doi.org/10. 1007/BF02055756
8. Wigner E (1932) On the quantum correction for thermodynamic equilibrium. Phys Rev 40(5):749–759. https://link.aps.org/doi/10.1103/PhysRev.40.749
9. Karl M, Gasenzer T (2017) Strongly anomalous non-thermal fixed point in a quenched two-dimensional Bose gas. New J Phys, 19(9):093014. https://doi.org/10.1088%2F1367-2630 %2Faa7eeb
10. Nicklas E, Karl M, Höfer M, Johnson A, Muessel W, Strobel H, Tomkovič J, Gasenzer T, Oberthaler MK (2015) Observation of scaling in the dynamics of a strongly quenched quantum gas. Phys Rev Lett 115:245301. https://link.aps.org/doi/10.1103/PhysRevLett.115.245301
11. Karl Markus, Nowak Boris, Gasenzer Thomas (2013) Universal scaling at nonthermal fixed points of a two-component Bose gas. Phys Rev A 88:063615. https://link.aps.org/doi/10.1103/ PhysRevA.88.063615
12. Wootters WK (1987) A Wigner-function formulation of finite-state quantum mechanics. Ann Phys, 176(1):1–21. http://www.sciencedirect.com/science/article/pii/000349168790176X
13. Wootters WK (2003) Picturing qubits in phase space. arXiv:quant-ph/0306135. http://arxiv. org/abs/quant-ph/0306135
14. Schachenmayer J, Pikovski A, Rey AM (2015) Many-body quantum spin dynamics with Monte Carlo trajectories on a discrete phase space. Phys Rev X 5:011022. https://link.aps.org/doi/10. 1103/PhysRevB.93.174302
15. Pucci Lorenzo, Roy Analabha, Kastner Michael (2016) Simulation of quantum spin dynamics by phase space sampling of Bogoliubov-Born-Green-Kirkwood-Yvon trajectories. Phys Rev B 93(17):174302. https://link.aps.org/doi/10.1103/PhysRevB.93.174302
16. Schachenmayer J, Pikovski A, Rey AM (2015) Dynamics of correlations in two-dimensional quantum spin models with long-range interactions: a phase-space Monte-Carlo study. New J Phys 17(6):065009. https://doi.org/10.1088/1367-2630/17/6/065009
17. Piñeiro Orioli A, Safavi-Naini A, Wall ML, Rey AM (2017) Nonequilibrium dynamics of spin-boson models from phase-space methods. Phys Rev A 96:033607. https://link.aps.org/doi/10. 1103/PhysRevA.96.033607
18. Michael Bonitz. *Quantum Kinetic Theory*. Springer International Publishing, 2015. https:// books.google.de/books?id=wW7_CgAAQBAJ
19. Paškauskas Rytis, Kastner Michael (2012) Equilibration in long-range quantum spin systems from a BBGKY perspective. J Stat Mech: Theory Exp 2012(02):P02005. https://doi.org/10. 1088%2F1742-5468%2F2012%2F02%2Fp02005
20. Pfeuty P (1970) The one-dimensional Ising model with a transverse field. Ann Phys (NY) 57:79–90. https://doi.org/10.1016/0003-4916(70)90270-8
21. Calabrese P, Essler FHL, Fagotti M (2012) Quantum quench in the transverse field Ising chain: I. time evolution of order parameter correlators. J Stat Mech: Theory Exp 2012(07):P07016. https://doi.org/10.1088%2F1742-5468%2F2012%2F07%2Fp07016
22. Calabrese P, Essler FHL, Fagotti M (2012) Quantum quenches in the transverse field Ising chain: II. stationary state properties. J Stat Mech: Theory Exp 2012(07):P07022. https://doi. org/10.1088%2F1742-5468%2F2012%2F07%2Fp07022
23. Lieb Elliott, Schultz Theodore, Mattis Daniel (1961) Two soluble models of an antiferro-magnetic chain. Ann Phys 16(3):407–466. http://www.sciencedirect.com/science/article/pii/ 0003491661901154
24. Hairer Ernst, Nørsett Syvert P, Wanner Gerhard (1993) Solving ordinary differential Equations I: nonstiff problems. Springer, Berlin. https://doi.org/10.1007/978-3-540-78862-1
25. Caflisch Russel E (1998) Monte Carlo and quasi-Monte Carlo methods. Acta Numer 7:1–49. https://doi.org/10.1017/S0962492900002804
26. Ryzhov Anton V, Yaffe Laurence G (2000) Large N quantum time evolution beyond leading order. Phys Rev D 62:125003. https://doi.org/10.1103/PhysRevD.62.125003
27. Berges Jürgen (2004) n-particle irreducible effective action techniques for Gauge theories. Phys Rev D 70:105010. https://doi.org/10.1103/PhysRevD.70.105010

28. Cakir H (2015) Dynamics of the transverse field Ising chain after a sudden quench. Master's thesis, Ruprecht-Karls-Universität Heidelberg
29. Karl Markus, Cakir Halil, Halimeh Jad C, Oberthaler Markus K, Kastner Michael, Gasenzer Thomas (2017) Universal equilibrium scaling functions at short times after a quench. Phys Rev E 96:022110. https://link.aps.org/doi/10.1103/PhysRevE.96.022110
30. Schmitt Markus, Heyl Markus (2018) Quantum dynamics in transverse-field Ising models from classical networks. SciPost Phys 4:013. https://scipost.org/10.21468/SciPostPhys.4.2.013
31. Fagotti M, Essler FHL (2013) Reduced density matrix after a quantum quench. Phys Rev B 87:245107. https://link.aps.org/doi/10.1103/PhysRevB.87.245107

# Chapter 5
# RBM-Based Wave Function Parametrization

Having benchmarked the discrete truncated Wigner approximation in the quantum critical regime, we now consider the performance of an ansatz based on quantum Monte Carlo methods to approximately simulate the same quenches in the transverse-field Ising model. Those methods are based on sampling states from a probability distribution underlying the Hilbert space to approximately evaluate expectation values of quantum operators. The task is then to efficiently find an expression representing this probability distribution. This suggests a connection to generative artificial neural network models which can be trained to represent probability distributions underlying given data. From those distributions, samples in the input space can be drawn. An approach using a restricted Boltzmann machine to parametrize quantum state vectors and apply Monte Carlo sampling of those states has been introduced in [1], where representations of ground states and dynamics are found via a variational ansatz.

In this chapter we recapitulate this ansatz in detail and benchmark its representational power for ground states and dynamics after sudden quenches in the transverse-field Ising model, with and without longitudinal field. We end the chapter with pointing out the limitations of the method.

This chapter contains discussions and results based on [2].

## 5.1 RBM Parametrization Ansatz

### 5.1.1 Quantum State Representation

Quantum Monte Carlo methods are a common approach to approximately simulate quantum many-body systems in an efficient manner. They are based on using Monte Carlo techniques to sample basis states from a probability distribution under-

© The Editor(s) (if applicable) and The Author(s), under exclusive
license to Springer Nature Switzerland AG 2020
S. Czischek, *Neural-Network Simulation of Strongly Correlated Quantum Systems*,
Springer Theses, https://doi.org/10.1007/978-3-030-52715-0_5

lying the Hilbert space and, with this, approximate expectation values of quantum operators [3].

The underlying probability distribution is defined via the basis-expansion coefficients $c_\mathbf{v}$ of the state vector,

$$|\Psi\rangle = \sum_{\{\mathbf{v}\}} c_\mathbf{v} |\mathbf{v}\rangle, \tag{5.1}$$

with basis states $|\mathbf{v}\rangle = |v_1, \ldots, v_N\rangle$ for an $N$-particle system. The sum runs over all $2^N$ basis states. The coefficients are generally complex, $c_\mathbf{v} \in \mathbb{C}$, and their squares $|c_\mathbf{v}|^2$ define probabilities for the basis states $\mathbf{v}$ [4–6]. This can be seen when considering the expectation value of an operator $O^{\text{diag}}$, which is diagonal in the considered basis,

$$
\begin{aligned}
\langle O^{\text{diag}} \rangle &= \langle \Psi | O^{\text{diag}} | \Psi \rangle \\
&= \sum_{\{\mathbf{v}\}} \sum_{\{\tilde{\mathbf{v}}\}} \langle \tilde{\mathbf{v}} | O^{\text{diag}} | \mathbf{v} \rangle c_\mathbf{v} c_{\tilde{\mathbf{v}}}^* \\
&= \sum_{\{\mathbf{v}\}} \sum_{\{\tilde{\mathbf{v}}\}} O^{\text{diag}}(\mathbf{v}) \, \delta_{\mathbf{v},\tilde{\mathbf{v}}} c_\mathbf{v} c_{\tilde{\mathbf{v}}}^* \\
&= \sum_{\{\mathbf{v}\}} O^{\text{diag}}(\mathbf{v}) |c_\mathbf{v}|^2 \\
&\approx \frac{1}{Q} \sum_{q=1}^{Q} O^{\text{diag}}(\mathbf{v}_q).
\end{aligned}
\tag{5.2}
$$

In the last line the sum over all basis states weighted by the probabilities $|c_\mathbf{v}|^2$ is approximated by $Q$ samples $\mathbf{v}_q$ drawn from this distribution. The approximation error is given by the variance.

To make these approximate simulations applicable, an efficient way to find and express the underlying probability distribution is necessary, as generally the full distribution is not known. Thus, a suitable parametrization of the squared basis-expansion coefficients, depending on the basis state $\mathbf{v}$, needs to be found. Several possible parametrization schemes exist and it depends on the problem of interest which of them is a wise choice.

The task of finding a suitable parametrization suggests a connection to generative artificial neural networks, such as the restricted Boltzmann machine (RBM) introduced in Sect. 3.2.1. These models can be trained to represent a probability distribution underlying some input data, where the explicit target distribution does not need to be known. Since a general RBM uses binary variables as neurons, parametrizing squared basis-expansion coefficients of quantum spin-1/2 systems is straightforward.

A general RBM defines a non-normalized Boltzmann distribution $P(\mathbf{v}, \mathbf{h}; \mathcal{W})$ underlying the $N$ visible and $M$ hidden neurons (or variables) $\mathbf{v}$ and $\mathbf{h}$, respectively. It depends on the weights and biases $\mathcal{W} = (\mathbf{d}, \mathbf{b}, W)$, as discussed in Sect. 3.2.1,

$$P(\mathbf{v}, \mathbf{h}; \mathcal{W}) = \exp\left[\sum_{i=1}^{N} d_i v_i + \sum_{j=1}^{M} b_j h_j + \sum_{i=1}^{N}\sum_{j=1}^{M} v_i W_{i,j} h_j\right]. \qquad (5.3)$$

Thus, the squared basis-expansion coefficients of a spin-1/2 system can be parametrized by an RBM,

$$|c_{\mathbf{v}}(\mathcal{W})|^2 = \sum_{\{\mathbf{h}\}} P(\mathbf{v}, \mathbf{h}; \mathcal{W}), \qquad (5.4)$$

with the sum running over all $2^M$ configurations of hidden variables. The visible variables are then identified with the spins in the system.

The sum over exponentially many hidden variable configurations can be approximated efficiently by sampling the hidden neurons from the underlying distribution and summing over those samples. This can be done in an efficient way via standard RBM methods, such as the block Gibbs sampling, which also provides samples for the visible variables according to $|c_{\mathbf{v}}(\mathcal{W})|^2$, see Sect. 3.2.2. Thus, the RBM can be trained such that it represents the probability distribution underlying the basis states in a certain quantum state. Expectation values of diagonal operators in this basis can then be evaluated efficiently in an approximate manner according to Eq. (5.2) by performing block Gibbs sampling to create the samples $\mathbf{v}_q$.

However, since the basis-expansion coefficients $c_{\mathbf{v}}$ are in general complex quantities, parametrizing the squared coefficients with an RBM does not capture the complex phases and hence does not fully represent the quantum state. Due to this, only expectation values of operators which are diagonal in the corresponding basis can be approximated with the parametrization of the squared coefficients. For general operators $O$, we need to include the complex phases,

$$\langle O \rangle = \sum_{\{\mathbf{v}\}}\sum_{\{\tilde{\mathbf{v}}\}} \langle \tilde{\mathbf{v}} | O | \mathbf{v} \rangle c_{\tilde{\mathbf{v}}}^* c_{\mathbf{v}}. \qquad (5.5)$$

This expectation value cannot be approximated via sampling from a real-valued probability distribution. In the most relevant cases, non-diagonal operators can still be evaluated efficiently, as we discuss in Sect. 5.1.3, but an explicit expression needs to be given for the complex coefficients $c_{\mathbf{v}}$.

To overcome the problem of missing complex phases in the parametrization, several ways to include them have been introduced [1, 7, 8]. In the following we consider the ansatz of using complex weights and biases in the RBM to directly parametrize the complex basis-expansion coefficients [1],

$$c_{\mathbf{v}}(\mathcal{W}) = \sum_{\{\mathbf{h}\}} \exp\left[\sum_{i=1}^{N} v_i d_i + \sum_{j=1}^{M} h_j b_j + \sum_{i=1}^{N}\sum_{j=1}^{M} v_i W_{i,j} h_j\right], \qquad (5.6)$$

where now $\mathcal{W} \in \mathbb{C}^{N+M+NM}$.

In contrast to a real-valued RBM as discussed in Sect. 3.2.1, the exponential of the network energy in the complex-valued RBM does not define a probability distribution over the visible and hidden variables anymore, but it rather yields a complex quantity. Only its square after summing over the hidden variables, $|c_{\mathbf{v}}(\mathcal{W})|^2$, gives a probability distribution from which spin configurations can be sampled [1]. Thus, there is no probability distribution defined for the hidden variables, but the sum over all of their configurations can be evaluated efficiently in an RBM,

$$
\begin{aligned}
c_{\mathbf{v}}(\mathcal{W}) &= \exp\left[\sum_{i=1}^{N} v_i d_i\right] \prod_{j=1}^{M} \left(\sum_{h_j=\pm 1} \exp\left[h_j b_j + \sum_{i=1}^{N} v_i W_{i,j} h_j\right]\right) \\
&= \exp\left[\sum_{i=1}^{N} v_i d_i\right] \prod_{j=1}^{M} 2\cosh\left[b_j + \sum_{i=1}^{N} v_i W_{i,j}\right].
\end{aligned}
\tag{5.7}
$$

With this expression, the probabilities $|c_{\mathbf{v}}(\mathcal{W})|^2$ can be evaluated efficiently and visible variables can be sampled. Additionally, an expression for the complex coefficients $c_{\mathbf{v}}(\mathcal{W})$ is still given.

The weights and biases in the RBM are free parameters and need to be adapted to represent a certain quantum state, which can be done with a variational ansatz. This corresponds to the training procedure in the real-valued RBM. However, new learning rules need to be defined for the complex-valued network, which we discuss in Sect. 5.1.2 [1].

The number of hidden variables is a hyper-parameter which can be chosen appropriately. As in a real-valued RBM, a larger number of hidden neurons increases the representational power. In contrast to a real-valued RBM, we cannot make the complex-valued network deeper by adding hidden layers, since there is no rule to sample the hidden variables. The efficient summation as in Eq. (5.7) is only possible for a single hidden layer.

With this, an RBM-based parametrization of quantum spin-1/2 state vectors has been introduced, on which Monte Carlo methods can be applied to approximately evaluate expectation values of quantum operators. Using a variational ansatz, the complex-valued RBM can be trained to represent ground states or even unitary dynamics in spin systems given some Hamiltonian $H$ [1].

## 5.1.2  Finding Ground States and Unitary Time Evolution

To find the right complex weights and biases representing certain ground and unitarily evolving states parametrized with a complex-valued RBM in the form of Eq. (5.6), a variational ansatz has been proposed in [1]. This ansatz can be accomplished by using a stochastic reconfiguration procedure, which can be interpreted as an unsupervised learning scheme [9, 10]. We first introduce the training scheme and then discuss in Sect. 5.1.3 how it can be applied efficiently.

In a variational ansatz the free parameters, which are here the weights and biases $\mathcal{W}$ in the RBM, are updated iteratively via some small variations $\delta\mathcal{W}$,

$$\mathcal{W}_\lambda^{p+1} = \mathcal{W}_\lambda^p + \varepsilon\delta\mathcal{W}_\lambda. \tag{5.8}$$

The step size $\varepsilon$ here corresponds to the learning rate. The index $p$ counts the iterations, and the index $\lambda$ refers to one element of $\mathcal{W}$, which can be a connecting weight or a bias.

Thus, given a state vector $|\Psi(\mathcal{W})\rangle$ parametrized with the RBM ansatz,

$$|\Psi(\mathcal{W})\rangle = \sum_{\{\mathbf{v}\}} c_\mathbf{v}(\mathcal{W})|\mathbf{v}\rangle, \tag{5.9}$$

it can be iteratively updated according to the variations in the weights. We can update the state vector via

$$\begin{aligned}|\Psi'(\mathcal{W})\rangle &= \exp\left[\Delta\sum_{\lambda=1}^{K}\delta\mathcal{W}_\lambda\frac{\partial}{\partial\mathcal{W}_\lambda}\right]|\Psi(\mathcal{W})\rangle \\ &= \left(\mathbb{1}+\Delta\sum_{\lambda=1}^{K}\delta\mathcal{W}_\lambda\frac{\partial}{\partial\mathcal{W}_\lambda}\right)|\Psi(\mathcal{W})\rangle + O\left(\Delta^2\right),\end{aligned} \tag{5.10}$$

with a small update step size $\Delta$, which is not necessarily related to $\varepsilon$. The sum runs over all $K = N + M + NM$ variational parameters in the RBM and $|\Psi'(\mathcal{W})\rangle$ is the new state vector [9, 10].

To find an expression for the variational updates we define a projected state $|\Psi_P(\mathcal{W})\rangle$ by applying a small imaginary time step of size $\Delta$ with Hamiltonian $H$ on the given state $|\Psi(\mathcal{W})\rangle$,

$$\begin{aligned}|\Psi_P(\mathcal{W})\rangle &= \exp\left[-\Delta H\right]|\Psi(\mathcal{W})\rangle \\ &= (\mathbb{1}-\Delta H)|\Psi(\mathcal{W})\rangle + O\left(\Delta^2\right).\end{aligned} \tag{5.11}$$

This imaginary time evolution applies a unitary time evolution operator solving the Schrödinger equation, analogously to Eq. (2.41) for the case of real time evolution. The time variable $t$ is set to an imaginary time variable $\tau = it$. With this replacement the evolution is expected to converge to the ground state of the system, as eigenstates with large eigenenergies are suppressed during the evolution. If we choose the step size $\Delta$ small enough, truncating at first order provides a valid approximation. The variations $\delta\mathcal{W}$ can thus be derived by minimizing the deviation between $|\Psi_P(\mathcal{W})\rangle$ and $|\Psi'(\mathcal{W})\rangle$, as the update then drives the state into the direction of the imaginary time evolution and a convergence to the ground state is expected for the parametrized state vector [9, 10].

For convenience sake we introduce the variational derivatives $\mathfrak{D}_\lambda$,

$$\mathfrak{D}_\lambda (\mathbf{v}) := \langle \mathbf{v} | \mathfrak{D}_\lambda | \mathbf{v} \rangle$$

$$= \frac{\partial}{\partial W_\lambda} \ln \left[ \langle \mathbf{v} | \Psi (W) \rangle \right] \tag{5.12}$$

$$= \frac{\partial}{\partial W_\lambda} \ln \left[ c_{\mathbf{v}} (W) \right] ,$$

so that the updated state vector, Eq. (5.10), can be expressed as [9, 10]

$$|\Psi' (W) \rangle = \left[ \mathbb{1} + \Delta \sum_{\lambda=1}^{K} \delta W_\lambda \mathfrak{D}_\lambda \right] |\Psi (W) \rangle + O \left( \Delta^2 \right) . \tag{5.13}$$

Considering the RBM parametrization of the basis-expansion coefficients, Eq. (5.6), the variational derivatives are given by [9, 10]

$$\mathfrak{D}_{d_i} (\mathbf{v}) = \frac{1}{c_{\mathbf{v}} (W)} \frac{\partial}{\partial d_i} c_{\mathbf{v}} (W) = v_i ,$$

$$\mathfrak{D}_{b_j} (\mathbf{v}) = \frac{1}{c_{\mathbf{v}} (W)} \frac{\partial}{\partial b_j} c_{\mathbf{v}} (W) = \tanh \left[ b_j + \sum_{i=1}^{N} v_i W_{i,j} \right] , \tag{5.14}$$

$$\mathfrak{D}_{W_{i,j}} (\mathbf{v}) = \frac{1}{c_{\mathbf{v}} (W)} \frac{\partial}{\partial W_{i,j}} c_{\mathbf{v}} (W) = v_i \tanh \left[ b_j + \sum_{l=1}^{N} v_l W_{l,j} \right] .$$

To minimize the deviation of the updated state vector from the projected state vector, we can maximize the overlap between the two,

$$\frac{\langle \Psi' (W) | \Psi_P (W) \rangle \langle \Psi_P (W) | \Psi' (W) \rangle}{\langle \Psi' (W) | \Psi' (W) \rangle \langle \Psi_P (W) | \Psi_P (W) \rangle} \overset{!}{=} 1 . \tag{5.15}$$

Solving for the variations $\delta W_\lambda$ to first order in $\Delta$ yields

$$\sum_{\mu=1}^{K} \delta W_\mu \left( \langle \mathfrak{D}_\lambda^* \mathfrak{D}_\mu \rangle - \langle \mathfrak{D}_\lambda \rangle \langle \mathfrak{D}_\mu^* \rangle \right) = - \langle \mathfrak{D}_\lambda^* H \rangle + \langle \mathfrak{D}_\lambda^* \rangle \langle H \rangle \quad \forall \lambda = 1, \ldots, K .$$

$$\tag{5.16}$$

The expectation values are calculated using the RBM parametrization of the wave function,

$$\langle O \rangle = \frac{\langle \Psi (W) | O | \Psi (W) \rangle}{\langle \Psi (W) | \Psi (W) \rangle}$$

$$= \frac{\sum_{\{\mathbf{v}\}} \sum_{\{\tilde{\mathbf{v}}\}} c_{\tilde{\mathbf{v}}}^* (W) c_{\mathbf{v}} (W) \langle \tilde{\mathbf{v}} | O | \mathbf{v} \rangle}{\sum_{\{\mathbf{v}\}} |c_{\mathbf{v}} (W)|^2} . \tag{5.17}$$

The denominator is necessary as the RBM-parametrized state vectors are generally not normalized. Equation (5.16) can be simplified by introducing a covariance matrix $S$ and a force vector $F$ with entries

$$S_{\lambda,\mu} = \langle \mathcal{D}_\lambda^* \mathcal{D}_\mu \rangle - \langle \mathcal{D}_\lambda^* \rangle \langle \mathcal{D}_\mu \rangle,$$
$$F_\lambda = \langle \mathcal{D}_\lambda^* H \rangle - \langle \mathcal{D}_\lambda^* \rangle \langle H \rangle. \tag{5.18}$$

Considering the vector $\delta W$ containing the variations of all weights and biases,

$$\delta W = \begin{pmatrix} \delta W_1 \\ \vdots \\ \delta W_K \end{pmatrix}, \tag{5.19}$$

the weight update in Eq. (5.16) can be expressed as

$$S\delta W = \begin{pmatrix} \sum_{\lambda=1}^{K} \delta W_\lambda S_{\lambda,0} \\ \vdots \\ \sum_{\lambda=1}^{K} \delta W_\lambda S_{\lambda,K} \end{pmatrix} = -F. \tag{5.20}$$

This can be solved for the variations in the weights,

$$\delta W = -S^{-1} F. \tag{5.21}$$

The update rule for the weights is thus

$$W^{p+1} = W^p - \varepsilon S^{-1} F, \tag{5.22}$$

which adjusts the variational parameters such that they follow the imaginary time evolution and converge to the ground state representation [1].

The update rule for the weights and biases to calculate a unitary time evolution can be derived in an analogous way, resulting in a scheme similar to time-dependent variational Monte Carlo approaches [11–13]. Here we define the projected state by applying the real time evolution,

$$\begin{aligned} |\Psi_P^t (W)\rangle &= \exp\left[-i\Delta H\right] |\Psi^t (W)\rangle \\ &= (\mathbb{1} - i\Delta H) |\Psi^t (W)\rangle + O\left(\Delta^2\right), \end{aligned} \tag{5.23}$$

where we again choose a small step size $\Delta$, which is now a real time step. The updated state is given using the derivative of the now time-dependent weights and biases,

$$|\Psi''(\mathcal{W})\rangle = \exp\left[i\Delta\frac{\partial}{\partial t}\right]|\Psi'(\mathcal{W})\rangle$$

$$= \left(\mathbb{1} + \Delta\frac{\partial}{\partial t}\right)|\Psi'(\mathcal{W})\rangle + O\left(\Delta^2\right) \tag{5.24}$$

$$= \left(\mathbb{1} + \Delta\sum_{\lambda=1}^{K}\frac{\partial\mathcal{W}_\lambda}{\partial t}\mathfrak{D}_\lambda\right)|\Psi'(\mathcal{W})\rangle + O\left(\Delta^2\right).$$

The variational derivatives as defined in Eq. (5.12) have been plugged in. Again, for small $\Delta$ a truncation at first order provides a valid approximation. By maximizing the overlap, as in Eq. (5.15), we arrive at an expression analogous to Eq. (5.21),

$$\frac{\partial\mathcal{W}}{\partial t} = -iS^{-1}F. \tag{5.25}$$

The covariance matrix $S$ and the force vector $F$ are defined as in Eq. (5.18). Note that the only difference to the update equation for the imaginary time evolution, Eq. (5.21), is a factor $i$. This shows the analogy between the imaginary time evolution to find the ground state and the unitary real time evolution [1].

The covariance matrix $S$ does not necessarily have full rank and is hence not necessarily invertible. To calculate the inverse for the weight updates, two methods can be used [1]. First, a regularization can be applied, which adds a diagonal matrix with infinitesimal entries to $S$. It then has full rank but does not change significantly, so that the updates are still well approximated. The regularization is defined as

$$S_{\lambda,\mu}^{\text{reg}} = S_{\lambda,\mu} + \kappa(p)\delta_{\lambda,\mu}S_{\lambda,\mu}, \tag{5.26}$$

where $\kappa(p) = \max(\kappa_0 m^p, \kappa_{\min})$ is a function starting at a small value $\kappa_0$ and decaying exponentially to a minimum value $\kappa_{\min}$ according to $m^p$, $m < 1$. Here $p$ counts the iterative weight-update steps. The choice of $\kappa_0$, $\kappa_{\min}$ and $m$ depends on the model considered and hence needs to be specified for different simulations [1].

The second ansatz is to calculate the Moore–Penrose pseudo-inverse instead of the exact inverse of $S$. This pseudo-inverse is calculated by applying a singular-value decomposition, $S = U\Sigma V^*$, with unitary matrices $U$, $V$ and a diagonal matrix $\Sigma$ containing the singular values. The inverse is given as $S^{-1} = V\Sigma^{-1}U^*$. If $S$ does not have full rank, neither does $\Sigma$. The pseudo-inverse of $\Sigma$ can be calculated by inverting the non-zero entries while leaving the rest unchanged. Undoing the decomposition yields the pseudo-inverse of $S$ [1].

In our simulations we find that it is best to apply both methods when calculating $S^{-1}$ to get a stable result. The stability depends on the choice of the parameters in the regularization with good choices varying for different spin models.

### 5.1.3 Evaluating Expectation Values

To calculate the covariance matrix and the force vector appearing in the weight updates in Eqs. (5.21) and (5.25), expectation values of the variational derivatives and of the Hamiltonian are needed. We also consider expectation values when measuring general observables. For a general operator $O$, these are given by

$$\langle O \rangle = \frac{\langle \Psi | O | \Psi \rangle}{\langle \Psi | \Psi \rangle} = \frac{\sum_{\{\mathbf{v}\},\{\tilde{\mathbf{v}}\}} c_\mathbf{v}^* \left( W \right) c_{\tilde{\mathbf{v}}} \left( W \right) \langle \mathbf{v} | O | \tilde{\mathbf{v}} \rangle}{\sum_{\{\mathbf{v}\}} |c_\mathbf{v} \left( W \right)|^2}. \tag{5.27}$$

The denominator compensates for the fact that the state vectors parametrized in the complex-valued RBM are generally not normalized. The sums run over all possible configurations and hence the numerator includes $2^N \times 2^N$ terms for an $N$-site spin-1/2 system. For large $N$, this sum cannot be calculated explicitly anymore due to the exponentially growing number of operations. Moreover, no efficient way to rewrite the sums exists.

Therefore, the expectation values need to be approximated in a quantum Monte Carlo fashion by not summing over all possible configurations of the spin system. Instead the sum runs only over a subset containing those configurations with large contributions to the expectation value, while those with small contributions can be neglected [1]. To create such a subset, Monte Carlo sampling is used. We are given the unnormalized probability distribution over the visible variables, $P(\mathbf{v}; W) = |c_\mathbf{v}(W)|^2$, from the complex-valued RBM. From this, spin configurations can be sampled via the Metropolis–Hastings scheme, which is a basic importance sampling method we introduce in detail in Sect. 5.1.5 [1].

Diagonal operators in the considered spin basis (here the $z$-basis), such as the variational derivatives $\mathfrak{D}_\lambda$, can then be evaluated approximately via

$$\langle \mathfrak{D}_\lambda \rangle = \frac{\sum_{\{\mathbf{v}\}} |c_\mathbf{v} \left( W \right)|^2 \, \mathfrak{D}_\lambda \left( \mathbf{v} \right)}{\sum_{\{\mathbf{v}\}} |c_\mathbf{v} \left( W \right)|^2}$$
$$\approx \frac{1}{Q} \sum_{q=1}^{Q} \mathfrak{D}_\lambda \left( \mathbf{v}_q \right). \tag{5.28}$$

Here we use $\langle \mathbf{v} | \tilde{\mathbf{v}} \rangle = \delta_{\mathbf{v}, \tilde{\mathbf{v}}}$ and draw $Q$ configurations $\mathbf{v}_q$ of visible variables from the probability distribution $|c_\mathbf{v}(W)|^2$. Expectation values of general observables $O$, which are not necessarily diagonal, can be approximated by [1]

$$\langle O \rangle = \frac{\sum_{\{\mathbf{v}\}} \sum_{\{\tilde{\mathbf{v}}\}} |c_\mathbf{v} \left( W \right)|^2 \frac{c_{\tilde{\mathbf{v}}}(W)}{c_\mathbf{v}(W)} \langle \mathbf{v} | O | \tilde{\mathbf{v}} \rangle}{\sum_{\{\mathbf{v}\}} |c_\mathbf{v} \left( W \right)|^2}$$
$$= \frac{\sum_{\{\mathbf{v}\}} |c_\mathbf{v} \left( W \right)|^2 \, O^{\mathrm{loc}} \left( \mathbf{v} \right)}{\sum_{\{\mathbf{v}\}} |c_\mathbf{v} \left( W \right)|^2} \tag{5.29}$$

$$\approx \frac{1}{Q} \sum_{q=1}^{Q} O^{\text{loc}} \left( \mathbf{v}_q \right).$$

$Q$ configurations $\mathbf{v}_q$ are sampled from $|c_{\mathbf{v}}(\mathcal{W})|^2$ and we introduce local observables

$$O^{\text{loc}} (\mathbf{v}) = \sum_{\{\tilde{\mathbf{v}}\}} \frac{c_{\tilde{\mathbf{v}}} (\mathcal{W})}{c_{\mathbf{v}} (\mathcal{W})} \langle \mathbf{v} |O| \tilde{\mathbf{v}} \rangle. \tag{5.30}$$

For diagonal operators this reduces to the expression in Eq. (5.28). If it is known for which states $\tilde{\mathbf{v}}$ the expression $\langle \mathbf{v}|O|\tilde{\mathbf{v}} \rangle$ is non-zero, one can sum up the contributions for these states instead of summing over all configurations in Eq. (5.30). This is more efficient if the operator $O$ is sparse, so that $\langle \tilde{\mathbf{v}}|O|\mathbf{v} \rangle$ vanishes for all but a polynomially scaling number of matrix elements for increasing system sizes. This property is given for the most commonly considered operators in quantum physics.

The approximation of expectation values via local observables and Monte Carlo sampling provides an efficient way to calculate the weight updates and evaluate the action of quantum operators on the represented spin system for large system sizes. However, the basis-expansion coefficients $c_{\mathbf{v}}(\mathcal{W})$ still need to be evaluated explicitly, which can here be done via calculating the network energy of the RBM according to Eq. (5.6) [1].

### 5.1.4  Including Translation Invariance

Having introduced the basics of the RBM parametrization of quantum spin-1/2 state vectors and the quantum Monte Carlo methods to evaluate expectation values of quantum operators, we discuss how this ansatz can be adapted to make it more efficient. Symmetries in spin systems can be used to reduce the number of degrees of freedom and thus make simulations computationally cheaper. For a spin chain with periodic boundary conditions translation invariance is often found, as it is the case in the transverse-field Ising model (TFIM). This means that the spins can be shifted around the ring without changing the system behavior.

These symmetries can also be implemented in the RBM parametrization of the state vector, resulting in certain weights and biases being equal. By determining these variational parameters and setting them equal explicitly, the RBM is forced to satisfy the symmetry and the number of variational parameters is reduced. This results in a more efficient representation [1, 14].

To include the translation invariance into the parametrization, we use a transformation-invariant RBM as defined in [14]. The transformation is here given by a translation of the visible variables by some shift $s$ along the ring, so that $N$ possible transformations can be applied. Thus, the hidden variables can be grouped into sets of $N$ neurons each, where the $j$th hidden variable compensates a shift by $j$

sites along the ring. The network energy can then be expressed as

$$
E\left(\mathbf{v}, \mathbf{h}; \mathcal{W}\right) = -\sum_{s=0}^{N-1}\left(\sum_{i=1}^{N} v_{(i+s)\bmod N} d_i\right) - \sum_{j=1}^{\alpha}\sum_{s=0}^{N-1} b_j h_{j,s}
$$
$$
-\sum_{j=1}^{\alpha}\sum_{s=0}^{N-1}\left(\sum_{i=1}^{N} v_{(i+s)\bmod N} W_{i,j} h_{j,s}\right)
$$
$$
= -d\sum_{i=1}^{N} v_i - \sum_{j=1}^{\alpha}\sum_{s=0}^{N-1} b_j h_{j,s} - \sum_{j=1}^{\alpha}\sum_{s=0}^{N-1}\left(\sum_{i=1}^{N} v_{(i+s)\bmod N} W_{i,j} h_{j,s}\right),
$$

$$(5.31)$$

with $d := \sum_{i=1}^{N} d_i$. Here we have $\alpha$ sets of $N$ hidden variables each, $M = N\alpha$, where $\alpha$ can be chosen freely to set the number of variational parameters. The first index of $h_{j,s}$ corresponds to the set the hidden variable is in, while the second index corresponds to the number of sites the spins are shifted around the circle.

Figure 5.1 visualizes the setup, where one can see that the visible biases are all equal. Furthermore, the hidden biases are equal within the sets of $N$ hidden variables and the connecting weights only depend on the distance between the connected visible and hidden variables and on the set of hidden variables they belong to [1, 14]. With this energy, the RBM parametrization of a state vector in a translationally invariant spin system is given by the coefficients [1]

$$
c_{\mathbf{v}}\left(\mathcal{W}\right) = \sum_{\{\mathbf{h}\}}\exp\left[d\sum_{i=1}^{N} v_i + \sum_{j=1}^{\alpha}\sum_{s=0}^{N-1} b_j h_{j,s} + \sum_{j=1}^{\alpha}\sum_{s=0}^{N-1}\sum_{i=1}^{N} v_{(i+s)\bmod N} W_{i,j} h_{j,s}\right]
$$

$$(5.32)$$

$$
= \exp\left[d\sum_{i=1}^{N} v_i\right]\prod_{j=1}^{\alpha}\prod_{s=0}^{N-1} 2\cosh\left[b_j + \sum_{i=1}^{N} v_{(i+s)\bmod N} W_{i,j}\right]. \qquad (5.33)
$$

The variational derivatives as defined in Eq. (5.12) then take the form

$$
\mathfrak{D}_d\left(\mathbf{v}\right) = \sum_{i=1}^{N} v_i,
$$
$$
\mathfrak{D}_{b_j}\left(\mathbf{v}\right) = \sum_{s=0}^{N-1}\tanh\left[b_j + \sum_{i=1}^{N} v_{(i+s)\bmod N} W_{i,j}\right], \qquad (5.34)
$$
$$
\mathfrak{D}_{W_{i,j}}\left(\mathbf{v}\right) = \sum_{s=0}^{N-1} v_{(i+s)\bmod N}\tanh\left[b_j + \sum_{l=1}^{N} v_{(l+s)\bmod N} W_{l,j}\right].
$$

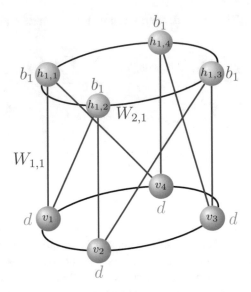

**Fig. 5.1** Visualization of the translation-invariant RBM for $N = 4$ visible and $M = 4$ hidden neurons. The network represents a spin chain with periodic boundary conditions. Same colors denote equal values for the variational parameters. The ring of hidden variables denotes the set of $N$ neurons in the hidden layer. For convenience sake only $\alpha = M/N = 1$ is shown here, a larger $\alpha$ would add more rings of hidden variables in an analogous way. Not all connecting weights are shown for better visibility

With this, the network can be trained to represent ground states or unitary dynamics just as introduced in Sect. 5.1.2. Moreover, expectation values of operators can be approximated in the same way as introduced in Sect. 5.1.3. Using $M = \alpha N$ hidden variables this symmetrization ansatz reduces the number of variational parameters from $N + M + MN$ in the parametrization as stated in Eq. (5.6) to $1 + \alpha + \alpha N = 1 + M/N + M$ in the translation-invariant parametrization.

The translation invariance is also represented by certain weights taking approximately equal values when finding the corresponding parameters variationally in the general parametrization. Thus, both representations are expected to give the same accuracy for the same number of hidden variables. Including the translation invariance hence makes the parametrization more efficient and less parameters need to be determined.

### 5.1.5  Metropolis–Hastings Sampling

A common approach to sample from a general probability distribution is the Metropolis–Hastings sampling, which is a Markov chain Monte Carlo method. This implies that a chain of samples is created where each sample depends only on the previous one [1, 15]. In the limit of long sampling times, or equivalently in the limit of many samples drawn, this chain of configurations moves to elements with large probabilities in the state space. These are the states which have large contributions to expectation values. To sample configurations of binary variables such as states in a spin-1/2 system, $v_i \in \{\pm 1\}$, from the underlying distribution $P(\mathbf{v})$, the procedure of the Metropolis–Hastings sampling consists of the following steps [15, 16],

1. start with a random spin configuration $\mathbf{v}$,
2. flip the variable at a random position $i$ in configuration $\mathbf{v}$ to create a new configuration $\tilde{\mathbf{v}}$, with $\tilde{v}_i = -v_i$, $\tilde{v}_j = v_j$ for $j \neq i$,
3. accept the new configuration with probability $A(\mathbf{v}, \tilde{\mathbf{v}}) = \min[1, P(\tilde{\mathbf{v}})/P(\mathbf{v})]$,

   - if the new configuration is accepted: $\tilde{\mathbf{v}} \to \mathbf{v}$,
   - else: $\mathbf{v} \to \mathbf{v}$,

4. add $\mathbf{v}$ to the set of sampled states and repeat the procedure from step 2. until the sample set reaches the desired size.

This scheme starts at a random point in the state space and for a large number of samples converges to a regime of states with large $P(\mathbf{v})$. This way, states with large contributions to the expectation values are sampled more often than states with smaller contributions, providing a weighting of the configurations according to $P(\mathbf{v})$. Here $P(\mathbf{v})$ does not need to be normalized, as only the ratio of two probabilities is considered [1, 15, 16].

To make the sampling procedure more efficient, the first samples can be neglected and the set creation starts only after $Q_i$ steps, such that the starting point is not totally random but already in a regime with large $P(\mathbf{v})$. Additionally, only every $Q_n$th sample can be added to the sample set to reduce the correlations between the individual samples and make them more independently distributed. The parameters $Q_i$ and $Q_n$ are hyper-parameters and can be chosen appropriately for the considered model, which is also true for the size $Q$ of the sample set [1, 15, 16].

This sampling scheme can be applied in the RBM parametrization to draw spin configurations from $P(\mathbf{v}; \mathcal{W}) = |c_\mathbf{v}(\mathcal{W})|^2$. As the probability distribution used in the Metropolis–Hastings sampling does not need to be normalized, the sampling from the unnormalized distribution represented by the RBM can be implemented straightforwardly in an efficient way.

## 5.2 Simulating Sudden Quenches in the TFIM

### 5.2.1 Setup of the Parametrization

To benchmark the complex RBM parametrization of ground state wave functions and dynamics in spin-1/2 systems, we consider sudden quenches in the TFIM and compare the simulation results with the exact solutions as stated in Sect. 2.5 [17–19].

To simulate the dynamics after the quenches, we first need to find the initial ground state parametrization via imaginary time evolution.

Given the TFIM Hamiltonian

$$H_{\text{TFIM}} = - J \sum_{i=1}^{N} \sigma_i^z \sigma_{(i+1)\text{mod}N}^z - h_{\text{t}} \sum_{i=1}^{N} \sigma_i^x, \tag{5.35}$$

the update rules for the weights to find the ground state can be derived using Eq. (5.22),

$$\begin{aligned} \mathcal{W}^{p+1} &= \mathcal{W}^p - \varepsilon S^{-1} F, \\ S_{\lambda,\mu} &= \langle \mathfrak{D}_\lambda^* \mathfrak{D}_\mu \rangle - \langle \mathfrak{D}_\lambda^* \rangle \langle \mathfrak{D}_\mu \rangle, \\ F_\lambda &= \langle \mathfrak{D}_\lambda^* H_{\text{TFIM}} \rangle - \langle \mathfrak{D}_\lambda^* \rangle \langle H_{\text{TFIM}} \rangle. \end{aligned} \tag{5.36}$$

Here we have rotated the system, so that the interaction in the TFIM is in the $z$-basis and the transverse field is applied in the $x$-basis. This can be done without loss of generality and provides the same analytical results.

As $\sigma^x$ operators appear in the Hamiltonian, we need to consider local operators when calculating energy expectation values,

$$\langle H_{\text{TFIM}} \rangle = \sum_{\{\mathbf{v}\}} E_{\text{loc}}(\mathbf{v}) \, |c_{\mathbf{v}}(\mathcal{W})|^2, \tag{5.37}$$

$$E_{\text{loc}}(\mathbf{v}) = \sum_{\{\tilde{\mathbf{v}}\}} \frac{c_{\tilde{\mathbf{v}}(\mathcal{W})}}{c_{\mathbf{v}}(\mathcal{W})} \langle \mathbf{v} | H_{\text{TFIM}} | \tilde{\mathbf{v}} \rangle. \tag{5.38}$$

The variational derivatives are diagonal, they can be calculated as stated in Eq. (5.34) and their expectation values can be evaluated according to Eq. (5.28). In the Hamiltonian we apply periodic boundary conditions, so that the model is invariant under translation. We can straightforwardly include this invariance into the RBM parametrization to reduce the number of parameters per hidden variable, as discussed in Sect. 5.1.4 [1, 14].

The same relations can be used to derive the update rules for the weights to learn the dynamics after a sudden quench of the transverse field, $h_{\text{t,i}} \to h_{\text{t,f}}$. The update rules are given by integrating Eq. (5.25) numerically using the (direct) Euler integration scheme, yielding

$$\mathcal{W}_{t+\Delta t} = \mathcal{W}_t - i \Delta t S^{-1} F. \tag{5.39}$$

To simulate the dynamics, we first find the ground state parametrization using the Hamiltonian with the initial transverse field $h_{\text{t,i}}$ and then start with these weights and biases to simulate the time evolution using the final transverse field $h_{\text{t,f}}$. Each update step corresponds to one time step [1].

In the end, we calculate correlation functions in the $z$-basis, $C_d^{zz}(t, h_{\text{t,f}})$, and compare them with the exact solutions. As these operators are diagonal, we can directly calculate the expectation values and do not need to consider local observables,

$$C_d^{zz}\left(t, h_{\mathrm{t,f}}\right) = \langle \sigma_0^z\left(t\right) \sigma_d^z\left(t\right) \rangle$$
$$= \sum_{\{\mathbf{v}\}} v_0\left(t\right) v_d\left(t\right) \left|c_{\mathbf{v}}\left(\mathcal{W}\right)\right|^2 . \tag{5.40}$$

Here we make use of the translation invariance in the TFIM, due to which we can choose one of the considered spins at site 0. The correlation function only depends on the relative distance between the two considered spins and not on their absolute positions.

In the following we consider two cases, one with a small system size and one with a longer spin chain. In the first case we can calculate the expectation values by explicitly summing over all possible spin states and we do not need the Monte Carlo sampling. We then have to weight each term with the probability $\left|c_{\mathbf{v}}\left(\mathcal{W}\right)\right|^2$ defined by the weights in the RBM. This way the expectation values can be calculated exactly. Deviations from the exact solution can then only result from the limited representational power of the RBM approach and from the accuracy of the weights therein. Thus, we can benchmark the ansatz itself.

In the second case we choose a large spin chain, so that we cannot sum over all spin configurations. Instead we use Metropolis–Hastings sampling, as introduced in Sect. 5.1.5, and approximate the expectation values of general operators $O$ via

$$\langle O \rangle \approx \frac{1}{Q} \sum_{q=1}^{Q} O^{\mathrm{loc}}\left(\mathbf{v}_q\right),$$
$$O^{\mathrm{loc}}\left(\mathbf{v}_q\right) = \sum_{\{\mathbf{v}\}} \frac{c_{\mathbf{v}}\left(\mathcal{W}\right)}{c_{\mathbf{v}_q}\left(\mathcal{W}\right)} \langle \mathbf{v}_q \left|O\right| \mathbf{v} \rangle. \tag{5.41}$$

$Q$ is the number of samples drawn from $\left|c_{\mathbf{v}}(\mathcal{W})\right|^2$ and $\mathbf{v}_q$ is the $q$th sampled configuration. This sampling procedure needs to be applied analogously when calculating the expectation value of the local energy and of the variational derivatives in the weight update procedure. By comparing the two cases of small and large system sizes, we can see how much the sampling affects the accuracy of the simulation method.

## 5.2.2 Finding Ground States

We first consider finding the ground state in the RBM-based parametrization and analyze the representational power by comparing the minimum energy reached after convergence in the optimization scheme with the exact ground state energy in the TFIM.

The exact ground state energy per spin-site can be calculated analytically via

$$E_0^{\text{TFIM}}(h_t, J) = -\frac{1}{N} \sum_{p \in P_N} \sqrt{J^2 + h_t^2 - 2Jh_t \cos[p]},$$

$$P_N = \frac{2\pi}{N} \left\{ -\frac{N-1}{2}, -\frac{N-3}{2}, \dots, \frac{N-1}{2} \right\}, \tag{5.42}$$

as discussed in Sect. 2.5.1 [17, 20]. We fix the energy scale by setting $J = 1$ and consider ground states for different transverse fields. We choose points in the ferromagnetic phase ($h_t = 0.5$), directly on the quantum phase transition ($h_t = 1$), and deep in the paramagnetic phase ($h_t = 100$). The last choice is already the initial condition for the sudden quenches we study in the following, as we have checked that it accurately approximates a fully $x$-polarized state with state vector

$$|\Psi\rangle = \bigotimes_{i=1}^{N} \left[ \frac{1}{\sqrt{2}} (|\uparrow\rangle_i + |\downarrow\rangle_i) \right]. \tag{5.43}$$

Here $|\uparrow\rangle_i$ denotes the spin at site $i$ being in the up-state. This would be the ground state for an infinitely large transverse field, $h_t \to \infty$.

To find the weights representing the ground states at these three points in the RBM approach, we start with some randomly chosen initial weights, which we draw from a uniform distribution in the region $[-0.001, 0.001]$. We then apply the update rules in Eq. (5.36) to minimize the energy and converge to the ground state. For the step size (or learning rate) we choose $\varepsilon = 0.01$ for $h_t = 0.5$ and $h_t = 1$, and $\varepsilon = 0.0001$ for $h_t = 100$, which we find to provide stable results. To invert the covariance matrix $S$ in the update rule, we combine the regularization approach and the Moore–Penrose pseudo-inverse as discussed in Sect. 5.1.2. We find suitable parameters $m = 0.9$, $\kappa_0 = 100$ and $\kappa_{\min} = 10^{-4}$ for the regularization after trying different values.

Figure 5.2 shows the results for benchmarking the ground state representation. The three panels (a), (b) and (c) show the different transverse fields $h_t = 0.5$, $h_t = 1$ and $h_t = 100$, respectively. In each panel the upper row shows the relative deviation from the exact energy defined as

$$R_E(h_t) = \left| \frac{E_0^{\text{RBM}}(h_t) - E_0^{\text{TFIM}}(h_t)}{E_0^{\text{TFIM}}(h_t)} \right|, \tag{5.44}$$

where $E_0^{\text{TFIM}}(h_t)$ is the exact ground state energy given by Eq. (5.42) and $E_0^{\text{RBM}}(h_t)$ is the energy resulting from the RBM parametrization. The lower row of the panels in Fig. 5.2 shows the simulated energy $E_0^{\text{RBM}}(h_t)$ as a function of iteration steps together with the exact solution $E_0^{\text{TFIM}}(h_t)$. Different colors denote different values of $\alpha = M/N$. With larger $\alpha$ the number $M$ of hidden variables and with this the number of variational parameters is increased.

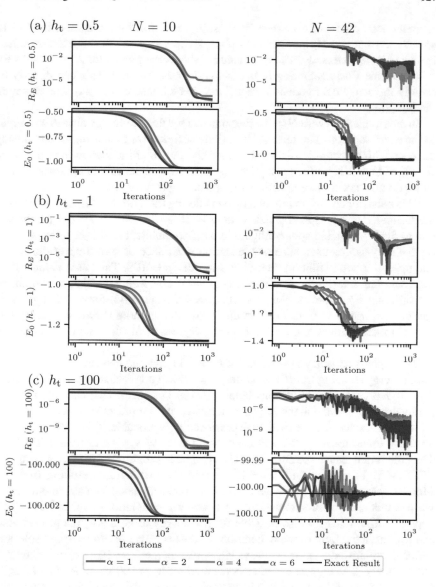

**Fig. 5.2** Ground state search in the RBM parametrization of the TFIM with $h_t = 0.5$ (**a**), $h_t = 1$ (**b**) and $h_t = 100$ (**c**). The upper row of each panel shows the relative deviation of the simulated energy from the exact value [Eq. (5.44)], while the lower row shows the simulated and exact energy [Eq. (5.42)] together, visualizing the convergence to the exact value. Left columns show results in a system with $N = 10$ sites and right columns show outcomes for $N = 42$ sites as functions of the iteration steps. Different colors correspond to results for different numbers of hidden variables, $\alpha = M/N$. For $N = 42$, Monte Carlo sampling needs to be applied to calculate expectation values

In the left column of each panel the results are shown for a system with $N = 10$ spin sites, where no Monte Carlo sampling is necessary and the expectation values can be calculated exactly. The right column shows the results for $N = 42$ sites for which Monte Carlo sampling in the evaluations is required. This can directly be seen in the statistical fluctuations appearing, which result from the approximation via sampling.

In the simulations with $N = 10$ sites one can see that the energy always converges to the exact solution. The relative deviations decrease for increasing $\alpha$, but already for $\alpha = 1$ high accuracy is reached, since the maximum deviation is found to be $R_E (h_t = 1) \approx 10^{-3}$. When going from $\alpha = 4$ to $\alpha = 6$ the gain in accuracy is small so we do not expect more benefits from increasing $\alpha$ further.

This behavior can be understood by considering the number of free parameters in the system. The exact spin-1/2 state with $N = 10$ sites has $2^N = 2^{10}$ free parameters due to the $2^N$ possible spin configurations. If we subtract those parameters which are fixed by symmetries, such as translation invariance or spin-flip invariance, we end up with a total Hilbert space dimension of $d_H = 108$. The RBM parametrization including translation invariance has $1 + \alpha + \alpha N$ weights which correspond to the variational parameters. For $N = 10$, $\alpha = 4$ this gives 45, for $\alpha = 6$ it gives 67 parameters, which is of the order of $d_H$. Thus, adding more variational parameters does not increase the accuracy much, as the degrees of freedom are mainly covered already.

The highest accuracy is reached for $h_t = 100$ and one can see that already the initial energy resulting from the randomly chosen weights is close to the exact energy. This is due to the initial weights being random but small. One can see from the parametrization form that the fully $x$-polarized state is represented if all weights are close to zero, since all states in the $z$-basis are equally probable. This justifies the high accuracy found for $h_t = 100$ and is also the reason why we rotate the TFIM here. Then the initial state to perform quenches can be approximated with high accuracy.

In the right column with $N = 42$ we find the fluctuations resulting from the Monte Carlo sampling, where we draw $Q = 1000$ samples. If not stated otherwise, we use this amount of samples for all upcoming simulations in this thesis. This approximation makes it also possible for the energy to go below the ground state energy, but in the limit of many iterations it always converges to the exact solution. While the relative deviation is always larger than in the $N = 10$ case, its behavior is still the same. The highest accuracy is reached for $h_t = 100$ while $R_E (h_t)$ is larger for $h_t = 0.5$ and $h_t = 1$, but is still around $R_E (h_t) \approx 10^{-3}$ and thus captures the exact energy well.

The effect of increasing $\alpha$ is much smaller than in the left column, for $h_t = 0.5$ and $h_t = 1$ more hidden variables even lead to larger relative deviations. Here the number of free parameters in the exact state is many orders of magnitude larger than the number of weights in the RBM parametrization. Thus, small changes in $\alpha$ are not significant compared to the large Hilbert space dimension. Slightly changing $\alpha$ does hence not have much influence and a much larger $\alpha$ is expected to be necessary to significantly increase the accuracy. This, however, is also computationally much more expensive. The larger relative deviations for increasing $\alpha$ result from the nonreproducibility of the calculations due to the Metropolis–Hastings sampling scheme. This means that the final energy still varies from run to run and the differences in the resulting energies are within these fluctuations.

In summary we can conclude that the ground states in the TFIM can be represented with sufficient accuracy in both the paramagnetic and the ferromagnetic phase and also directly at the quantum critical point, considering large as well as small system sizes. Expectation values are then calculated with and without Monte Carlo sampling, respectively. Especially for $h_t = 100$ the exact ground state can be represented very well for all values of $\alpha$, meaning that it can be used as an initial state to study dynamics after sudden quenches.

### 5.2.3  Sudden Quenches in Small Systems

We start with benchmarking the approximate simulation of dynamics using the RBM parametrization by considering sudden quenches in the TFIM with $N = 10$ spin sites. We do not need Monte Carlo sampling but can directly benchmark the representational power of the ansatz. We evaluate the correlation function,

$$C_d^{zz}\left(t, h_{t,f}\right) = \langle \sigma_0^z(t)\, \sigma_d^z(t) \rangle, \tag{5.45}$$

and compare the simulations with the exact results for quenches from an initial state deep in the paramagnetic phase, $h_{t,i} = 100$, to different final transverse fields $h_{t,f}$ [18, 19, 21].

Figure 5.3 shows the correlation function as a function of time after sudden quenches into the ferromagnetic regime [Fig. 5.3a], onto the quantum-critical point [Fig. 5.3b] and within the paramagnetic regime [Fig. 5.3c]. The RBM simulations for $\alpha = 1$, $\alpha = 2$, $\alpha = 4$ and $\alpha = 6$ are plotted together with the exact solution. For completeness also the results of the semi-classical discrete truncated Wigner approximation (dTWA) as discussed in Chap. 4 are shown. The correlations are calculated for three relative distances between the two spins, $d = 1$ corresponding to nearest-neighbor correlations (solid lines), $d = 2$ being next-to-nearest-neighbor correlations (dashed lines), and $d = 3$ (dotted lines).

For the quenches into the ferromagnetic and within the paramagnetic regime [panels (a) and (c)], the RBM ansatz works well and captures the exact solution accurately at all times, even for $\alpha = 1$. Considering the quench onto the quantum critical point [panel (b)], clear deviations from the exact solution appear for small $\alpha$, while convergence to the analytical curve is observed when increasing $\alpha$. Choosing $\alpha = 6$ the simulation follows the exact solution. As discussed already for the ground state search, for these $\alpha$ the number of variational parameters is of the order of the Hilbert space dimension after taking symmetries into account. Thus, it does not come unexpected that the exact solution is represented well, since almost all degrees of freedom are covered by the variational parameters. The dTWA results only capture

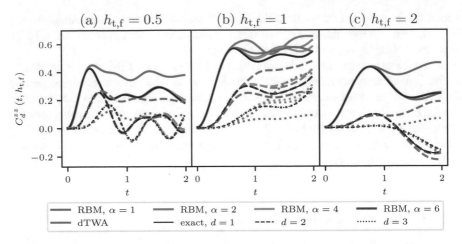

**Fig. 5.3** Correlation function $C_d^{zz}(t, h_{t,f})$ [Eq. (5.45)] as a function of time after a sudden quench to final fields $h_{t,f} = 0.5$ (**a**), $h_{t,f} = 1$ (**b**) and $h_{t,f} = 2$ (**c**). RBM simulations for different numbers of hidden variables $\alpha = M/N$ are compared to the exact solution for relative distances $d = 1$ (solid lines), $d = 2$ (dashed lines) and $d = 3$ (dotted lines) in a system with $N = 10$ sites. At the quantum critical point ($h_{t,f} = 1$) convergence is reached for $\alpha = 6$, where the variational parameters cover the Hilbert space dimension. For the other cases $\alpha = 1$ is sufficient to represent the exact dynamics. DTWA results are shown for completeness. Figure adapted from [2]

the exact solution well at short times. For later times clear deviations appear, while the RBM approach captures the exact solution at long and short times.

To analyze the performance of the RBM parametrization as a function of the final transverse field $h_{t,f}$, we look at the short-distance correlation length $\xi_1(t, h_{t,f})$ as introduced in Sect. 2.5.2 [18, 19, 21]. This correlation length can be extracted by fitting an exponential function to the correlation function at $d < 3$,

$$C_d^{zz}(t, h_{t,f}) \propto \exp\left[-\frac{d}{\xi_1(t, h_{t,f})}\right]. \tag{5.46}$$

We have already shown in Fig. 5.3 that the RBM parametrization captures the exact dynamics well for these small relative distances. Thus, it is reasonable to study the correlation length for quenches to different final fields for a more detailed benchmark.

Figure 5.4 shows the correlation length at a fixed time $t = 1$ after sudden quenches as a function of the final transverse field $h_{t,f}$. Again, the RBM representations for $\alpha = 1$, $\alpha = 2$, $\alpha = 4$ and $\alpha = 6$ are plotted together with the exact solution and the dTWA result.

Deviations can be found for almost all $h_{t,f}$ in the dTWA, except for very small $h_{t,f} \approx 0$, where also the RBM approach works well. In this regime the transverse field is a small perturbation to the classical Ising model, so that the system can be described by a perturbation approach [22]. The RBM ansatz captures the exact solution well in most cases. Already for $\alpha = 1$ the simulations follow the exact solution closely

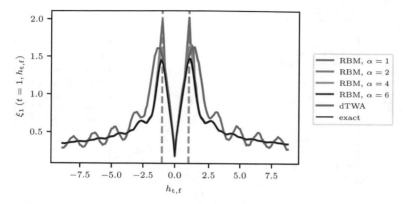

**Fig. 5.4** Correlation length $\xi_1(t, h_{t,f})$ [Eq. (5.46)] at a fixed time $t = 1$ after sudden quenches within the paramagnetic, into the ferromagnetic and across the ferromagnetic regime as a function of $h_{t,f}$. RBM simulations for different numbers of hidden variables $\alpha = M/N$ with $N = 10$ are compared to the exact solution and dTWA results. Dashed vertical lines denote the quantum critical points. At those the correlation length reaches its maximum and large $\alpha$ need to be chosen, while everywhere else $\alpha = 1$ describes the exact solution accurately. Figure adapted from [2]

everywhere except for the quantum critical points, which are indicated by dashed vertical lines. Around these points, at $h_{t,f} = 1$ and $h_{t,f} = -1$, convergence to the exact solution is found with increasing $\alpha$ and for $\alpha = 6$ the exact result is captured. This is what we have already observed when considering the correlation function. Here we can see that the RBM parametrization struggles at the maxima of the correlation length and larger $\alpha$ are necessary. This suggests that strong long-range correlations require more hidden variables to be captured.

Figure 5.5 shows the correlation length at a fixed time $t = 1$ after quenches within the paramagnetic regime as a function of the distance $\epsilon$,

$$\epsilon = \left| \frac{h_{t,f} - h_{t,c}}{h_{t,c}} \right|$$

$$= h_{t,f} - 1, \tag{5.47}$$

from the quantum critical point $h_{t,c} = 1$ on a logarithmic scale. As discussed in Sect. 2.5.2, for the quenches from an effectively fully polarized state we expect the behavior of the correlation length to be determined by a generalized Gibbs ensemble (GGE),

$$\xi_{\text{GGE}}(\epsilon) = \frac{1}{\ln(2\epsilon + 2)}. \tag{5.48}$$

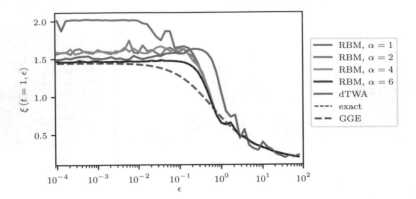

**Fig. 5.5** Correlation length $\xi_1(t, \epsilon)$ [Eq. (5.46)] at a fixed time $t = 1$ after sudden quenches within the paramagnetic regime as a function of the final distance $\epsilon$ [Eq. (5.47)] from the quantum critical point in a system with $N = 10$ sites. RBM simulations for different $\alpha$ are compared to the exact solution (dashed black line) and dTWA results. Convergence to the exact solution is found for $\alpha = 6$ in the regime of small $\epsilon$. The GGE behavior, to which the exact solution converges at late times in large systems, is plotted for completeness according to Eq. (5.48) (brown dashed line). Figure adapted from [2]

Thus, we expect it to saturate in the vicinity of the quantum critical point instead of diverging due to the quantum phase transition being only present for zero temperature [18, 19, 21]. To study if these effects are captured in the RBM parametrization the simulation results for different $\alpha$ are plotted together with the exact solution and the expected GGE behavior. DTWA results are shown for completeness. We find that the GGE behavior is not captured by the dTWA simulations. Again, $\alpha = 6$ is necessary in the RBM parametrization to capture the exact solution in the vicinity of the quantum critical point, but then the simulations follow the exact curve for all $\epsilon$.

At these short times even the exact solution does not show the GGE behavior, which is plotted for comparison by the brown dashed line in Fig. 5.5. For those small system sizes it is not even possible to find the GGE behavior at later times due to finite-size effects appearing [18, 19, 21]. For larger spin systems we expect the RBM parametrization to show the GGE behavior for large enough $\alpha$ given that the simulations work in the same way, since they follow the exact solution so well in small systems. When simulating longer spin chains, on the one hand Monte Carlo sampling is necessary to calculate observables, and on the other hand it is not clear how large $\alpha$ needs to be chosen to capture the exact solution.

The Hilbert space dimension $d_H$ scales exponentially with the system size and for $N = 10$ we find that the number of weights in the RBM parametrization needs to be of the order of $d_H$. To analyze how the necessary $\alpha$ scales with the system size, we consider the correlation function after a sudden quench onto the quantum critical

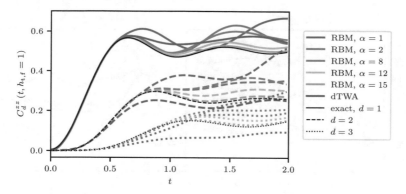

**Fig. 5.6** Correlation function $C_d^{zz}(t, h_{t,f})$ [Eq. (5.45)] as a function of time after a sudden quench to $h_{t,f} = 1$ onto the quantum critical point in a system with $N = 12$ sites. Results from the RBM parametrization for different $\alpha$ are compared to the exact solution and to dTWA calculations. The correlations are shown for relative distances $d = 1$ (solid lines), $d = 2$ (dashed lines) and $d = 3$ (dotted lines). To capture the exact solution, $\alpha = 15$ is necessary, where the variational parameters cover the Hilbert space dimension. This suggests an exponential scaling of the number of hidden neurons with the system size. Figure adapted from [2]

point in a system with $N = 12$ sites. We still do not need Monte Carlo sampling but can sum over all spin states to calculate expectation values.

Figure 5.6 shows the resulting correlation function for $d = 1$, $d = 2$ and $d = 3$ as a function of time, together with the exact solution and the dTWA. We consider different values of $\alpha$ and find deviations for small $\alpha$, as well as a convergence to the exact curve with increasing $\alpha$. However, we find that we need to choose larger values than in the previous case with $N = 10$ sites. We even have to choose $\alpha = 15$ to capture the exact solution with high accuracy, which corresponds to $1 + \alpha + M = 196$ variational parameters. This is again of the order of the Hilbert space dimension after symmetries have been subtracted, which is $d_H = 352$ for $N = 12$ sites. This shows that the required $\alpha$ does not scale linearly with the system size, suggesting that the Hilbert space dimension needs to be covered by the number of variational parameters. We have checked that, far from the quantum phase transition, $\alpha = 1$ is sufficient to capture the exact solution also for $N = 12$ sites.

### 5.2.4  Going to Longer Spin Chains

After benchmarking the representational power of the RBM parametrization on small spin chains, we can check how the Metropolis–Hastings sampling affects the results, as this is necessary to calculate expectation values for general large spin systems. To

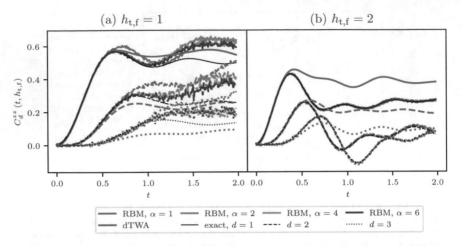

**Fig. 5.7** Correlation function $C_d^{zz}(t, h_{t,f})$ [Eq. (5.45)] as a function of time after a sudden quench onto the quantum critical point [**a**, $h_{t,f} = 1$] and within the paramagnetic regime [**b**, $h_{t,f} = 2$] in a system with $N = 42$ sites. Results of the RBM parametrization for different numbers of hidden variables are compared to the exact solution and dTWA results for relative distances $d = 1$ (solid lines), $d = 2$ (dashed lines) and $d = 3$ (dotted lines) between the sites. Convergence to the exact solution is not reachable at the quantum critical point, while away from it choosing $\alpha = 1$ is sufficient. Figure adapted from [2]

analyze this we consider the same quenches as for the small systems and study the dynamics of the correlation function in a spin chain with $N = 42$ sites.

Figure 5.7 shows the resulting dynamics for quenches to $h_{t,f} = 1$ [Fig. 5.7a] and $h_{t,f} = 2$ [Fig. 5.7b]. The correlation function is shown for three relative distances, $d = 1$ (solid lines), $d = 2$ (dashed lines) and $d = 3$ (dotted lines). The RBM results for $\alpha = 1, \alpha = 2, \alpha = 4$ and $\alpha = 6$ are plotted together with exact and dTWA results.

Compared to the simulations for small system sizes in Fig. 5.3, the RBM results show small fluctuations which result from the Monte Carlo sampling and also appear in the ground state search in Fig. 5.2. They can be reduced further by increasing the sample size, but then also the computations become more expensive.

Apart from that, we find the same behavior as for $N = 10$. For the quench to $h_{t,f} = 2, \alpha = 1$ is sufficient to capture the exact dynamics accurately while deviations appear for the quench onto the quantum critical point. Increasing $\alpha$ does not have much influence here. This is probably due to the fact that compared to the Hilbert space dimension the number of variational parameters is still very small, even for $\alpha = 6$, as discussed in Sect. 5.2.2 for the ground state search.

We have already observed for $N = 12$ in Fig. 5.6 that $\alpha$ does not scale linearly with the system size in the quantum critical regime, so here we expect a much larger $\alpha$ necessary to converge to the exact solution. This fact renders the RBM parametrization inefficient in the quantum critical regime, since the computational cost seems to grow exponentially with the system size. This is analogous to the exact calculations and hence limits the simulations to small system sizes.

## 5.3  Performance in the TFIM in a Longitudinal Field

### 5.3.1  Benchmarking Short-Chain Simulations

Having pointed out the limitation of the RBM parametrization to small system sizes for quenches into the quantum critical regime, we next benchmark it in a non-integrable model. Therefore we consider the TFIM in an additional longitudinal field $h_l$, a model we refer to as LTFIM,

$$H_{\text{LTFIM}} = -\sum_{i=1}^{N} \sigma_i^z \sigma_{(i+1)\text{mod}N}^z - h_t \sum_{i=1}^{N} \sigma_i^x - h_l \sum_{i=1}^{N} \sigma_i^z. \qquad (5.49)$$

We first simulate short spin-chains with $N = 10$ sites, where we can compare our simulation results with exact diagonalization outcomes.

In the following we consider dynamics after sudden quenches from an effectively fully $x$-polarized state, choosing $h_{t,i} = 100$ and $h_{l,i} = 0$ as initial fields. We then quench the system to different final transverse and longitudinal fields $h_{t,f}$ and $h_{l,f}$, respectively, as depicted in Fig. 5.8. For the longitudinal field the model is symmetric around $h_l = 0$, so that we only consider quenches to positive $h_l > 0$.

When simulating the dynamics of the correlation function after the quench, we find an exponential decay at short relative distances $d < 3$, so that we can extract a correlation length $\xi_l(t, h_{t,f}, h_{l,f})$ in the same way as in the TFIM case, see Eq. (5.46) [18, 19, 21]. Thus, in the following we study the dynamics of this correlation length. Figure 5.9 shows the dynamics in a color-plot as a function of time $t$ and final transverse field $h_{t,f}$ for quenches to longitudinal fields $h_{l,f} = 0$ [Fig. 5.9a], $h_{l,f} = 1$ [Fig. 5.9b], $h_{l,f} = 2$ [Fig. 5.9c] and $h_{l,f} = 3$ [Fig. 5.9d]. Each row consists of four plots, where the first plot shows the exact diagonalization result and the remaining

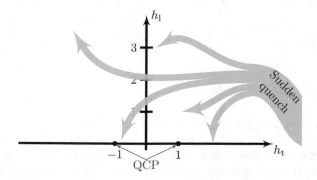

**Fig. 5.8** Illustration of sudden quenches in the TFIM with an additional longitudinal field $h_l$. The quantum phase transition only exists for $h_l = 0$ at the quantum critical points (QCP). We consider sudden quenches from a large transverse field and zero longitudinal field to different final values for both, ending up in the paramagnetic (large $h_t$, $h_l$) or ferromagnetic (small $h_t$) phase

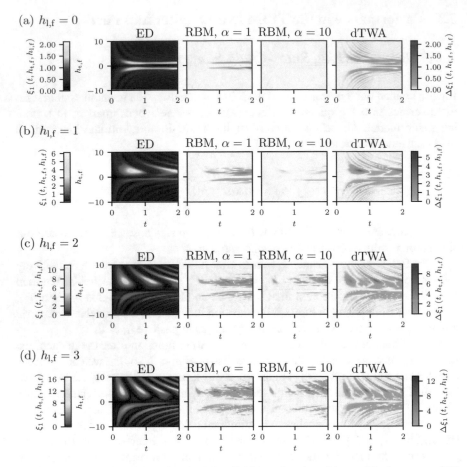

**Fig. 5.9** Correlation length $\xi_1(t, h_{t,f}, h_{1,f})$ [Eq. (5.46)] as a function of time and final transverse field $h_{t,f}$ after sudden quenches to different final longitudinal fields $h_{1,f} = 0$ (**a**), $h_{1,f} = 1$ (**b**), $h_{1,f} = 2$ (**c**) and $h_{1,f} = 3$ [(**d**)] in the LTFIM with $N = 10$ sites. Each row shows from left to right the exact solution calculated via exact diagonalization (ED) and absolute deviations from it [Eq. (5.50)] resulting from the RBM parametrization with $\alpha = 1$ and $\alpha = 10$ as well as from dTWA. Figure partly adapted from [2]

three show the absolute deviation,

$$\Delta\xi_1\left(t, h_{t,f}, h_{1,f}\right) = \left|\xi_1^{\text{exact}}\left(t, h_{t,f}, h_{1,f}\right) - \xi_1^{\text{sim}}\left(t, h_{t,f}, h_{1,f}\right)\right|, \tag{5.50}$$

of the simulated correlation length $\xi_1^{\text{sim}}(t, h_{t,f})$ from the exact solution $\xi_1^{\text{exact}}(t, h_{t,f})$. The different plots correspond to the RBM parametrization with $\alpha = 1$ and $\alpha = 10$, and to dTWA results, respectively, from left to right.

Figure 5.9a shows the result for the TFIM and one can clearly see the deviations at the quantum critical points in the RBM simulations for $\alpha = 1$. One can furthermore observe that these deviations vanish for $\alpha = 10$. The dTWA shows deviations in all regimes, but they get smaller further away from the quantum critical point, as already observed in Sects. 5.2.3 and 4.4. In Fig. 5.9b, c and d a longitudinal field $h_{1,f} > 0$ is present after the quench, but the same observations can be made. Even though there is no quantum phase transition in this model, we find regimes with maximum correlation lengths around $h_{1,f} \approx 0$. In these regions, deviations in the RBM parametrization for $\alpha = 1$ appear which get smaller for $\alpha = 10$, but do not vanish completely. For $\alpha = 10$ the number of variational parameters is $1 + \alpha + M = 111$ and exceeds the dimension of the Hilbert space, $d_H = 108$, after symmetrization.

The correlation lengths reach larger values with increasing longitudinal field and the regimes of deviations appearing in the RBM parametrization with $\alpha = 1$ grow, indicating a relation between the correlation length and the performance of the RBM approach. The dTWA shows deviations for all transverse fields and only works well at very short times.

In the exact diagonalization result we find the expected Rabi-oscillations, as discussed in Sect. 2.5.3, in the dynamics of the correlation length. These result from the interplay of the transverse and the longitudinal field. To analyze the oscillations in more detail, Fig. 5.10 has the same setup as Fig. 5.9. Instead of the absolute deviations, the correlation lengths resulting from the RBM parametrization with $\alpha = 1$ and $\alpha = 10$ and from the dTWA are plotted. One can see that the oscillations are captured well in the RBM simulations.

The green lines in the exact diagonalization plots in Fig. 5.10 show the contours of the corresponding von-Neumann entanglement entropy,

$$S_{vN} = - \text{Tr}\left[\rho_A \log \rho_A\right], \tag{5.51}$$

with $\rho_A$ being the reduced density matrix of one part in the bipartite system. The von-Neumann entanglement entropy is calculated via time-dependent density-matrix renormalization group (tDMRG) and is also plotted in Fig. 2.8. From the contours one can see that the entanglement entropy also gets large in the regime where deviations appear in the RBM approach. In these regions of large entanglement also methods based on matrix product states struggle, so that our results suggest that the RBM parametrization shows limitations in the same regime as other existing simulation methods.

## 5.3.2 Monte Carlo Sampling in Large Systems

Considering longer spin chains with $N = 42$ sites, we cannot use exact diagonalization anymore. We can, however, compare the RBM simulations with tDMRG results. Those are known to represent the exact solution with good accuracy, as discussed in Sect. 2.6 [23–28]. We analyze the performance of the RBM parametrization in a regime where the tDMRG method struggles and a large bond dimension $D$ is required to be chosen for convergence. This is the case for regimes of strong quantum entan-

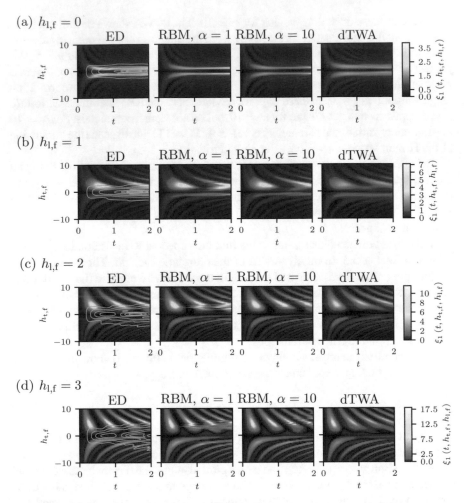

**Fig. 5.10** Correlation length $\xi_1(t, h_{t,f}, h_{1,f})$ [Eq. (5.46)] as a function of time and final transverse field $h_{t,f}$ after sudden quenches to final longitudinal fields $h_{1,f} = 0$ (**a**), $h_{1,f} = 1$ (**b**), $h_{1,f} = 2$ (**c**) and $h_{1,f} = 3$ (**d**) in the LTFIM with $N = 10$ sites. Each row shows the exact solution calculated via exact diagonalization (ED), as well as results from the RBM parametrization with $\alpha = 1$, $\alpha = 10$, and from dTWA simulations, from left to right. Green lines in the exact solution show contours of the von-Neumann entanglement entropy [Eq. (5.51)] as plotted in Fig. 2.8. Figure adapted from [2]

glement and hence we consider a sudden quench to $h_{t,f} = 0.5$, $h_{1,f} = 1$. We know from Fig. 2.8 that the von-Neumann entanglement entropy reaches large values in this regime [23–28]. From Fig. 5.9 we still expect the RBM ansatz to work well for these fields.

Figure 5.11a shows the resulting nearest-neighbor (solid lines) and next-to-nearest-neighbor (dashed lines) correlation functions depending on time $t$ after the quench. The RBM results for $\alpha = 1$ and $\alpha = 2$ are plotted together with tDMRG

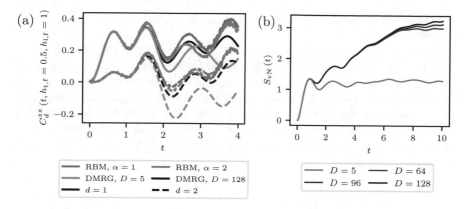

**Fig. 5.11** Correlation function $C_d^{zz}(t, h_{t,f}, h_{l,f})$ [Eq. (5.45)] as a function of time after a sudden quench to $(h_{t,f} = 0.5, h_{l,f} = 1)$ in the LTFIM with $N = 42$ sites (**a**). RBM simulations with $\alpha = 1$ and $\alpha = 2$ are compared to tDMRG results with bond dimensions $D = 5$ and $D = 128$. $D = 128$ approximates the exact solution with good accuracy for the times considered. The results for $D = 5$ are shown as the number of variational parameters is of the same order as in the RBM simulations. For this case both approximation methods show comparable behavior. Correlations are shown for relative distances $d = 1$ (solid lines) and $d = 2$ (dashed lines). Panel (**b**) shows the von-Neumann entanglement entropy [Eq. (5.51)] resulting from tDMRG calculations with different bond dimensions as functions of time after the same quench. Overlapping curves denote convergence of the approximation method. Figure adapted from [2]

results with bond dimensions $D = 5$ and $D = 128$. Figure 5.11b shows the von-Neumann entanglement entropy [Eq. (5.51)] calculated using tDMRG with different bond dimensions as a function of time after the quench. When the simulations have converged, nothing changes when increasing $D$ and the different lines lie on top of each other. This justifies the choice of $D = 128$ in Fig. 5.11a, as we see that up to the times considered the tDMRG has already converged for smaller bond-dimensions. Thus, we can interpret these calculations as exact solutions.

The calculation with $D = 5$ is shown since this bond dimension provides 50 variational parameters, which is of the order of the number of free parameters in the RBM parametrization with $\alpha = 1$ (44 variational parameters) and $\alpha = 2$ (87 variational parameters). Thus, these results are fairly comparable with the RBM results in Fig. 5.11a. In the comparison we find that deviations from the tDMRG calculation with $D = 128$ appear around the same times for both the RBM calculations and the tDMRG simulations with $D = 5$. This suggests that the RBM parametrization struggles in the same regime as existing methods based on matrix product states given the same number of variational parameters. As we find exponential scaling in the bond dimension of the tDMRG in regimes with strong entanglement, we also expect exponential scaling of $\alpha$ in the RBM approach there. However, the tDMRG can be implemented in a much more efficient way than the RBM parametrization, so that we cannot increase $\alpha$ as far as we can increase $D$ due to limited computational power. Thus, the numerical precision reachable by the RBM parametrization falls behind the results of existing methods based on matrix product states on comparable computational resources.

In summary we have made the same observations as in the TFIM also in the simulations of the LTFIM. In regimes of large correlation lengths and strong entanglement entropy the required number of hidden variables to reach a certain accuracy scales exponentially with the system size. Again, the whole Hilbert space dimension needs to be covered by the variational parameters, rendering the RBM ansatz inefficient and limiting it to small system sizes. In regimes of weak correlations, however, the method performs well and requires only a small amount of hidden variables, even for large system sizes.

## 5.4  Limitations of the RBM Parametrization

### 5.4.1  Representational Power

When benchmarking the RBM parametrization representing ground states and dynamics of quantum spin-1/2 systems, we have experienced the limits of this simulation ansatz. From the results shown so far, we have found that in regimes of large correlation lengths the method becomes inefficient, as exponentially many hidden variables are necessary to capture the dynamics accurately. To further analyze where these deviations result from, we can first check whether the limiting factor lies in the representational power of the parametrization ansatz or in the way the time evolution is calculated. To study this we use exact diagonalization to find the state $|\Psi(t)\rangle$ the system is in at a time $t$ after a sudden quench, and then find the weights representing this state in the RBM parametrization.

To be able to apply exact diagonalization we consider a small system with $N = 10$ sites and prepare it in a fully $x$-polarized initial state,

$$|\Psi_0\rangle = \bigotimes_{i=1}^{N} \left[ \frac{1}{\sqrt{2}} \left( |\uparrow\rangle_i + |\downarrow\rangle_i \right) \right],  \tag{5.52}$$

with $|\uparrow\rangle_i$ denoting the spin at site $i$ in the up-state. We then perform a sudden quench onto the quantum critical point $h_{t,f} = 1$ in the TFIM. We can see in Fig. 5.3 that for this quench $\alpha = 6$ is necessary in the RBM parametrization to capture the exact dynamics of the nearest-neighbor correlation function $C_{d=1}^{zz}(t, h_{t,f}) = \langle \sigma_0^z(t)\sigma_1^z(t)\rangle$ at times $t \geq 0.5$. Thus, in the following we consider the state at time $t = 1$ after the quench and analyze whether it can generally be represented by the RBM ansatz and how the accuracy scales with $\alpha$.

Exact diagonalization provides the basis expansion coefficients $c_v^{\text{ex}}$ of the exact state for each spin configuration,

$$|\Psi\,(t=1)\rangle = \exp\left[-iH\right]|\Psi_0\rangle$$
$$= \sum_{\{\mathbf{v}\}} c_{\mathbf{v}}^{\mathrm{ex}}\,(t=1)\,|\mathbf{v}\rangle. \tag{5.53}$$

To find the weights to represent this state in the RBM parametrization, we make use of the fact that the squared coefficients $|c_{\mathbf{v}}^{\mathrm{ex}}(t)|^2$ give the probability of the system to be in the corresponding state. Thus, we can consider the Kullback–Leibler divergence as stated in Eq. (3.31), which is a measure for the distance between two probability distributions. It hence should be minimized when considering the exact and the represented distributions. By defining the exact probability $P(\mathbf{v})$ and the unnormalized probability $P(\mathbf{v}; \mathcal{W})$ resulting from the RBM parametrization,

$$P\,(\mathbf{v}) = \left|c_{\mathbf{v}}^{\mathrm{ex}}\,(t=1)\right|^2, \tag{5.54}$$
$$P\,(\mathbf{v}; \mathcal{W}) = \left|c_{\mathbf{v}}\,(\mathcal{W})\right|^2, \tag{5.55}$$

the Kullback–Leibler divergence can be written as

$$
\begin{aligned}
\Xi\,(\mathcal{W}) :&- D_{\mathrm{KL}}\left[P\,(\mathbf{v})\,||\,\frac{P\,(\mathbf{v}; \mathcal{W})}{Z\,(\mathcal{W})}\right] \\
&= \sum_{\{\mathbf{v}\}} P\,(\mathbf{v}) \log\left[\frac{P\,(\mathbf{v})\,Z\,(\mathcal{W})}{P\,(\mathbf{v}; \mathcal{W})}\right],
\end{aligned} \tag{5.56}
$$

with the normalization factor $Z(\mathcal{W})$ for $P(\mathbf{v}; \mathcal{W})$,

$$Z\,(\mathcal{W}) = \sum_{\{\mathbf{v}\}} P\,(\mathbf{v}; \mathcal{W})\,. \tag{5.57}$$

The distribution $P(\mathbf{v})$ is normalized by construction.

To find weights in the RBM parametrization minimizing the Kullback–Leibler divergence we can use a gradient descent approach. The derivative of $\Xi(\mathcal{W})$ with respect to the individual weights is given by

$$\frac{\partial}{\partial \mathcal{W}_\lambda^\alpha} \Xi\,(\mathcal{W}) = \sum_{\{\mathbf{v}\}}\left[-\frac{P\,(\mathbf{v})}{P\,(\mathbf{v}; \mathcal{W})}\frac{\partial}{\partial \mathcal{W}_\lambda^\alpha}P\,(\mathbf{v}; \mathcal{W}) + \frac{P\,(\mathbf{v})}{Z\,(\mathcal{W})}\frac{\partial}{\partial \mathcal{W}_\lambda^\alpha}Z\,(\mathcal{W})\right], \tag{5.58}$$

where $\mathcal{W}_\lambda^\alpha$ denotes one element of the weight vector $\mathcal{W} = \mathcal{W}^{\mathrm{R}} + i\mathcal{W}^{\mathrm{I}}$, considering the real and imaginary parts individually, $\alpha \in \{\mathrm{R}, \mathrm{I}\}$. We can also express $P(\mathbf{v}; \mathcal{W})$ in terms of the real and imaginary parts of the weights,

$$P(\mathbf{v}; \mathcal{W}) = \exp\left[2\sum_{i=1}^{N} d_i^{\mathrm{R}} v_i\right] \prod_{j=1}^{M} 2\left(\cosh\left[2\sum_{i=1}^{N} W_{i,j}^{\mathrm{R}} v_i + 2b_j^{\mathrm{R}}\right]\right.$$

$$\left.\cos\left[2\sum_{i=1}^{N} W_{i,j}^{\mathrm{I}} v_i + 2b_j^{\mathrm{I}}\right]\right), \qquad (5.59)$$

where the derivatives with respect to the individual parts can be calculated. These can then be plugged into the update equation of the gradient descent ansatz, yielding

$$\mathcal{W}_\lambda^\alpha(p+1) = \mathcal{W}_\lambda^\alpha(p) - \varepsilon \sum_{\{\mathbf{v}\}} \left\{ \frac{P(\mathbf{v})}{P(\mathbf{v}, p; \mathcal{W})} \frac{\partial}{\partial \mathcal{W}_\lambda^\alpha} P(\mathbf{v}, p; \mathcal{W}) \right.$$

$$\left. - \frac{P(\mathbf{v})}{\sum_{\{\tilde{\mathbf{v}}\}} P(\tilde{\mathbf{v}}, p; \mathcal{W})} \sum_{\{\tilde{\mathbf{v}}\}} \frac{\partial}{\partial \mathcal{W}_\lambda^\alpha} P(\tilde{\mathbf{v}}, p; \mathcal{W}) \right\},$$

$$(5.60)$$

where $p$ is the iteration step and $\varepsilon$ is the step width, or learning rate.

For such small system sizes with $N = 10$ sites no sampling is necessary, so we always sum over the whole state space. Figure 5.12a shows the Kullback–Leibler divergence as a function of the total number of iterations for finding the RBM parametrization with different numbers of hidden variables, $\alpha = M/N$. All simulations are run five times and the results are averaged, with the shaded regions denoting statistical fluctuations. One can clearly see a convergence to zero for all $\alpha$. One moreover observes that for $\alpha = 1$ the Kullback–Leibler divergence saturates at a larger value than for $\alpha > 1$. Increasing the number of hidden variables further does not provide any improvement. This is not in accordance with the previous observations in Sect. 5.2.3, where we needed at least $\alpha = 6$ to capture the exact solution, while here $\alpha = 2$ appears to be sufficient.

In Fig. 5.12b the nearest-neighbor correlation resulting from the saturated probability distribution after $10^5$ training iterations is plotted as a function of $\alpha$ together with the exact solution in the inset. The main plot shows the absolute deviation between the two. The exact solution is calculated via exact diagonalization and error bars again denote statistical fluctuations of averaging five simulation runs. There one can see that the exact correlation is captured well for $\alpha = 1$, which is again in contradiction to the results in Sect. 5.2.3. However, the deviation still gets smaller with increasing $\alpha$, but saturates around $10^{-4}$. The larger value at $\alpha = 8$ is due to strong statistical fluctuations and is expected to vanish when averaging more simulation runs, as can be seen in the large error bars.

We have to keep in mind that we do not find the weights representing the full quantum state when minimizing the Kullback–Leibler divergence. Instead, we only find weights representing the exact amplitude, or probability distribution underlying the states. The phases of the coefficients $c_\mathbf{v}(\mathcal{W})$ can still differ from the exact solution. To also get the phases right, we would need to perform measurements in

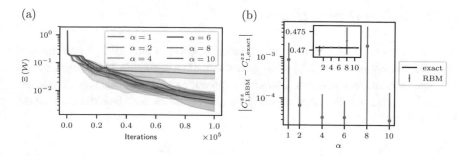

**Fig. 5.12** Convergence of the RBM parametrization to the probability distribution defined by the state at time $t = 1$ after a sudden quench onto the quantum critical point in the TFIM with $N = 10$ sites. Weight updates are done via minimizing the Kullback–Leibler divergence $\Xi(W)$ [Eq. (5.56)]. The evolution of the Kullback–Leibler divergence is shown in panel **a** as a function of the weight update iterations for parametrizations with different numbers of hidden variables $\alpha = M/N$. Panel **b** shows the nearest-neighbor correlation function $C_d^{zz}(t = 1, h_{t,f} = 1)$ [Eq. (5.45)] reached after $10^5$ iterations. The main plot shows the absolute deviation from the exact solution as a function of $\alpha$, while the inset shows the direct evaluation of the correlation on the $y$-axis with the $x$-axis being the same as in the main plot. The exact solution is shown for comparison in the inset. For all calculations five individual runs are made to find the weights and are averaged, where the shaded regions in **a** and the error bars in **b** denote the statistical fluctuations. The plots show that the probability distribution underlying the basis states can in principle be represented well by the RBM

all possible bases, meaning that each spin needs to be measured in the $x$-, $y$- and $z$-basis and all possible combinations of measurements are necessary. This exceeds our computational resources, so that we do not run simulations finding the weights to represent the full state.

Thus, from Fig. 5.12 we can conclude that in principle it is possible to efficiently represent a spin state with the right amplitude and the right $zz$-correlation at time $t = 1$ after a sudden quench onto the quantum critical point. Nevertheless, it is not clear whether also the phases can be captured and hence whether the full quantum state can be represented. Deviations in the phases might cause the deviations found in Sect. 5.2.3 and would explain the convergence with increasing $\alpha$.

On the other hand, it might also be possible that this state at time $t = 1$ after the quench can be represented with good accuracy for $\alpha = 1$, but a different state at an earlier time cannot. As during the simulation of the time evolution the weights are always updated depending on their previous values, a deviation from the exact solution at an early time can lead to deviations also appearing at later times.

Thus, from our simulations we cannot directly tell whether the deviations found in Sect. 5.2.3 result from the limited representational power of the network or from problems appearing in the way the dynamics are simulated. However, the convergence we find when increasing $\alpha$ motivates us to further analyze the representational power of the parametrization.

### 5.4.2 Adding Hidden Layers

A different ansatz to modify the representational power of the RBM parametrization
is to make the network deeper by adding another hidden layer with $\tilde{M}$ hidden variables
$\tilde{\mathbf{h}}$. These are connected to the variables in the already existing hidden layer via weights
$\tilde{W}$ and have biases $\tilde{\mathbf{b}}$. The network energy of this so-called deep Boltzmann machine
then reads

$$E\left(\mathbf{v}, \mathbf{h}, \tilde{\mathbf{h}}\right) = -\sum_{i=1}^{N}\sum_{j=1}^{M} v_i W_{i,j} h_j - \sum_{j=1}^{M}\sum_{l=1}^{\tilde{M}} h_j \tilde{W}_{j,l} \tilde{h}_l - \sum_{i=1}^{N} v_i d_i - \sum_{j=1}^{M} h_j b_j - \sum_{l=1}^{\tilde{M}} \tilde{h}_l \tilde{b}_l,$$

$$(5.61)$$

and the coefficients $c_{\mathbf{v}}(\mathcal{W})$ of the basis states are parametrized as

$$c_{\mathbf{v}}(\mathcal{W}) = \sum_{\{\mathbf{h}\}}\sum_{\{\tilde{\mathbf{h}}\}} \exp\left[-E\left(\mathbf{v}, \mathbf{h}, \tilde{\mathbf{h}}\right)\right], \qquad (5.62)$$

where we need to sum over both layers of hidden variables.

However, in contrast to the two-layer RBM considered so far this sum over both
layers of hidden variables cannot be performed analytically anymore for large spin
systems. Only the sum over one of the two layers can be rewritten as a product over
the hidden variables similar to Eq. (5.7), while the remaining sum over $2^M$ (or $2^{\tilde{M}}$)
terms needs to be calculated explicitly. Thus, we are limited to small numbers of
hidden variables, where we can still sum over all of them to perform simulations
with such a deep network.

To study the effect of the additional layer, we simulate sudden quenches from
an effectively fully $x$-polarized initial state onto the quantum critical point in the
TFIM, similar to Sect. 5.2.3. We consider the time evolution of the nearest-neighbor
$zz$-correlation function parametrized with a three-layer network in a system with
$N = 10$ spin sites. Choosing $M = \tilde{M} = 10$ hidden variables per layer, we can still
sum them out explicitly.

Figure 5.13 shows the result of the simulation, where we let the calculations run for
five times and average the outcome, yielding statistical fluctuations depicted by the
shaded regions. The exact solution is shown for comparison, as well as the results
of simulations using a two-layer RBM. We find that deviations in the three-layer
simulations appear already at earlier times than for the two-layer RBM. This comes
unexpected, since the representational power of the RBM should increase with an
additional layer [29].

We can assume here that the earlier breakdown results from the way the weights
are updated to describe the time evolution rather than from the representational power.
It is probably harder to train the deeper networks accordingly. Furthermore, small
deviations appearing at some point in time can be even more amplified at later times
than in the case of shallower networks. This would also explain the larger fluctuations

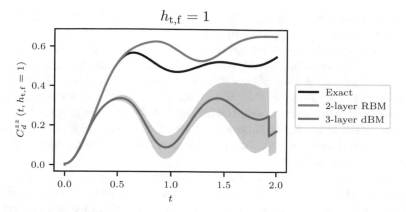

**Fig. 5.13** Simulation of the nearest-neighbor $zz$-correlation function [Eq. (5.45)] as a function of time after a sudden quench from an effectively fully $x$-polarized state onto the quantum critical point, $h_{\mathrm{t,f}} = 1$, in the TFIM with $N = 10$ sites. Simulations are done with a three-layer deep Boltzmann machine (dBM) using $M = \tilde{M} = 10$ hidden variables per layer (blue line). Outcomes of five runs are averaged, where the shaded regions denote statistical fluctuations. The exact solution is plotted for comparison (black line), as well as the results of simulations using a two-layer RBM with $M = 10$ hidden variables (orange line). Using the deeper network, deviations from the exact solution are found at earlier times than in the simulations with the two-layer RBM

found at later times after the quench. However, we can again not check where exactly these deviations result from. We can only conclude that deeper network structures are not necessarily helpful here.

## 5.5 Summary

In this chapter we have benchmarked quantum Monte Carlo simulations of quantum many-body systems based on a parametrization of the state vectors via a restricted Boltzmann machine (RBM) [1]. We have considered simulations of ground states and dynamics after sudden quenches in the transverse-field Ising model (TFIM). As the model is integrable, we have compared the simulations with exact solutions [18, 19]. We have first analyzed the convergence to ground state energies and afterwards considered sudden quenches from an initial state deep in the paramagnetic phase to different final transverse fields. For the dynamics we have also shown a direct comparison to discrete truncated Wigner approximation (dTWA) results for completeness, which led to the same observations as in Chap. 4.

When benchmarking the ground state search, we have considered states in both the paramagnetic and the ferromagnetic phase, as well as directly on the quantum critical point. We have shown that the exact ground state energy can be represented with good accuracy in all regimes, for small and large spin chains. For the latter case we have applied a Metropolis–Hastings sampling scheme. We have shown that the precision of the results is high in both cases.

To benchmark the representation of dynamics, we have considered small system sizes, where expectation values can be calculated by exactly summing over all spin configurations. We have shown that the efficiency of the method is limited for quenches into the vicinity of the quantum critical points. There the whole Hilbert space dimension needs to be covered by the number of network parameters to capture the exact solution accurately. However, far from the quantum phase transitions, the RBM parametrization captures the dynamics after quenches well, even for as many hidden as visible variables.

Going to larger spin systems and using Metropolis–Hastings sampling to calculate expectation values, we have shown that the necessary number of hidden variables does not scale linearly with the system size in the vicinity of the quantum critical point. Thus, the RBM parametrization does not perform efficiently in these regimes. Further away from the quantum phase transition we have found the exact dynamics to be captured well again, even for a small and linearly scaling number of hidden variables.

For a more general benchmark, we have added a longitudinal field to the TFIM and made the model non-integrable [30]. We have considered sudden quenches from an initial state with large transverse and zero longitudinal field to different final fields. We have again considered small and large system sizes with Metropolis–Hastings sampling for the latter case. We have compared the results with exact solutions obtained via exact diagonalization for the small system sizes or time-dependent density-matrix renormalization group (tDMRG) for the larger system sizes. We have shown that still a large number of hidden variables, exceeding the Hilbert space dimension, is necessary to capture the exact solution accurately in regimes of large correlation lengths. At regimes of smaller correlation lengths we have found the RBM parametrization to represent the exact solution accurately even with a small number of hidden variables.

We have further considered a sudden quench into a regime where we know that tDMRG methods struggle and need an exponentially increasing bond-dimension to converge to the exact solution [23–28]. We have pointed out that comparable results for the RBM parametrization and the tDMRG can be reached when using approximately equal numbers of variational parameters. Thus, the RBM approach seems to struggle in the same regime as existing methods based on matrix product states. However, the tDMRG can be implemented in a more efficient way. Given comparable computational resources the bond dimension in the tDMRG can hence be increased further than the number of hidden variables in the RBM parametrization. Thus, the RBM approach cannot compete with the numerical performance of methods based on matrix product states.

At the end of the chapter we have studied the deviations appearing in the simulations of sudden quenches in the TFIM with the RBM parametrization. We have checked whether they result from the limited representational power of the parametrization ansatz or rather from the way the weights are updated to simulate dynamics. In this context we have shown that adding another small hidden layer does not improve the performance of the method, but we were not able to make a conclusion about what causes the limitations.

In summary we have pointed out the struggling of the RBM parametrization ansatz for quenches into regimes of strong long-range correlations. This is also a problematic regime for other approximate simulation methods, such as tDMRG [24, 31] or dTWA, see Chap. 4. Nevertheless, the RBM method can also be applied in higher-dimensional systems, where a similar performance is expected and where existing simulation methods are scarce. Thus, the ansatz is expected to enable an efficient simulation of weakly-correlated spin systems in higher spatial dimensions.

# References

1. Carleo G, Troyer M (2017) Solving the quantum many-body problem with artificial neural networks. Science 355(6325):602–606. http://science.sciencemag.org/content/355/6325/602
2. Czischek S, Gärttner M, Gasenzer T (2018) Quenches near Ising quantum criticality as a challenge for artificial neural networks. Phys Rev B 98:024311. https://doi.org/10.1103/PhysRevB.98.024311
3. Rubenstein B (2017) Introduction to the variational Monte Carlo Method in quantum chemistry and physics. Springer, Singapore, pp 285–313. https://doi.org/10.1007/978-981-10-2502-0_10
4. Nielsen MA, Chuang IL (2010) Quantum Computation and Quantum Information: 10th, Anniversary Edition. Cambridge University Press, Cambridge. https://doi.org/10.1017/CBO9780511976667
5. Troyer M (2012) Lecture notes on computational quantum physics. http://edu.itp.phys.ethz.ch/fs12/cqp/
6. Bartelmann M, Lüst D, Wipf A, Rebhan A, Feuerbacher B, Krüger T (2015) Die Entstehung der Quantenphysik. Springer, Berlin. https://doi.org/10.1007/978-3-642-54618-1_21
7. Torlai G, Mazzola G, Carrasquilla J, Troyer M, Melko R, Carleo G (2018) Neural-network quantum state tomography. Nat Phys 14:447–450. https://doi.org/10.1038/s41567-018-0048-5
8. Torlai G, Melko RG (2018) Latent space purification via neural density operators. Phys Rev Lett 120:240503. https://link.aps.org/doi/10.1103/PhysRevLett.120.240503
9. Sorella S (2001) Generalized Lanczos algorithm for variational quantum Monte Carlo. Phys Rev B 64(2):024512. https://link.aps.org/doi/10.1103/PhysRevB.64.024512
10. Sorella S, Casula M, Rocca D (2007) Weak binding between two aromatic rings: feeling the van der Waals attraction by quantum Monte Carlo methods. J Chem Phys 127(1):014105. http://dx.doi.org/10.1063/1.2746035
11. Carleo G, Becca F, Schió M, Fabrizio M (2012) Localization and glassy dynamics of many-body quantum systems. Sci Rep 2(243). http://dx.doi.org/10.1038/srep00243
12. Carleo G, Becca F, Sanchez-Palencia L, Sorella S, Fabrizio M (2014) Light-cone effect and supersonic correlations in one- and two-dimensional bosonic superfluids. Phys Rev A 89(3):031602. https://link.aps.org/doi/10.1103/PhysRevA.89.031602
13. Ido K, Ohgoe T, Imada M (2015) Time-dependent many-variable variational Monte Carlo method for nonequilibrium strongly correlated electron systems. Phys Rev B 92(24):245106. https://link.aps.org/doi/10.1103/PhysRevB.92.245106
14. Sohn K, Lee H (2012) Learning invariant representations with local transformations. In: ICML'12. Omnipress, USA, pp 1339–1346. http://dl.acm.org/citation.cfm?id=3042573.3042745

15. Metropolis N, Rosenbluth AW, Rosenbluth MN, Teller AH, Teller E (1953) Equation of state calculations by fast computing machines. J Chem Phys 21(6):1087–1092. https://doi.org/10.1063/1.1699114
16. Bishop C (2006) Pattern recognition and machine learning. Springer, New York. https://www.springer.com/de/book/9780387310732?referer=www.springer.de
17. Pfeuty P (1970) The one-dimensional Ising model with a transverse field. Ann Phys (NY) 57:79–90. https://doi.org/10.1016/0003-4916(70)90270-8
18. Calabrese P, Essler FHL, Fagotti M (2012) Quantum quench in the transverse field Ising chain: I. time evolution of order parameter correlators. J Stat Mech: Theory Exp 2012(07):P07016. https://doi.org/10.1088%2F1742-5468%2F2012%2F07%2Fp07016
19. Calabrese P, Essler FHL, Fagotti M (2012) Quantum quenches in the transverse field Ising chain: II. stationary state properties. J Stat Mech: Theory Exp 2012(07):P07022. https://doi.org/10.1088%2F1742-5468%2F2012%2F07%2Fp07022
20. Lieb E, Schultz T, Mattis D (1961) Two soluble models of an antiferromagnetic chain. Ann Phys 16(3):407–466. http://www.sciencedirect.com/science/article/pii/0003491661901154
21. Karl M, Cakir H, Halimeh JC, Oberthaler MK, Kastner M, Gasenzer T (2017) Universal equilibrium scaling functions at short times after a quench. Phys Rev E 96:022110. https://link.aps.org/doi/10.1103/PhysRevE.96.022110
22. Schmitt M, Heyl M (2018) Quantum dynamics in transverse-field Ising models from classical networks. SciPost Phys 4:013. https://scipost.org/10.21468/SciPostPhys.4.2.013
23. White SR (1992) Density matrix formulation for quantum renormalization groups. Phys Rev Lett 69:2863–2866. https://link.aps.org/doi/10.1103/PhysRevLett.69.2863
24. Schollwöck U (2011) The density-matrix renormalization group in the age of matrix product states. Ann Phys 326(1):96–192. http://www.sciencedirect.com/science/article/pii/S0003491610001752
25. Vidal G (2004) Efficient simulation of one-dimensional quantum many-body systems. Phys Rev Lett 93:040502. https://link.aps.org/doi/10.1103/PhysRevLett.93.040502
26. Daley AJ, Kollath C, Schollwöck U, Vidal G (2004) Time-dependent density-matrix renormalization-group using adaptive effective Hilbert spaces. J Stat Mech: Theory Exp 2004(04):P04005. http://stacks.iop.org/1742-5468/2004/i=04/a=P04005
27. Sharma S, Suzuki S, Dutta A (2015) Quenches and dynamical phase transitions in a non-integrable quantum Ising model. Phys Rev B 92:104306. https://link.aps.org/doi/10.1103/PhysRevB.92.104306
28. Haegeman J, Lubich C, Oseledets I, Vandereycken B, Verstraete F (2016) Unifying time evolution and optimization with matrix product states. Phys Rev B 94:165116. https://link.aps.org/doi/10.1103/PhysRevB.94.165116
29. Gao X, Duan L-M (2017) Efficient representation of quantum many-body states with deep neural networks. Nat Commun 8(1). https://doi.org/10.1038/s41467-017-00705-2
30. Ovchinnikov AA, Dmitriev DV, Krivnov VY, Cheranovskii VO (2003) Antiferromagnetic Ising chain in a mixed transverse and longitudinal magnetic field. Phys Rev B 68(21):214406. https://link.aps.org/doi/10.1103/PhysRevB.68.214406
31. Bridgeman JC, Chubb CT (2017) Hand-waving and interpretive dance: an introductory course on tensor networks. J Phys A: Math Theor 50(22):223001. https://doi.org/10.1088%2F1751-8121%2Faa6dc3

# Part III
# Relating Quantum Systems and Neuromorphic Hardware

# Chapter 6
# Deep Neural Networks and Phase Reweighting

Having pointed out the limitations of approximate simulation methods for quantum spin-1/2 systems on classical computers, we are motivated to study whether other computing devices going beyond the von-Neumann architecture are more suitable for these tasks. One such device is the neuromorphic hardware present in the BrainScaleS group at Heidelberg University, which we introduced in Sect. 3.5. This hardware can efficiently produce samples from Boltzmann distributions underlying the visible and hidden variables in (restricted) Boltzmann machines [1–4]. Furthermore, we have introduced the parametrization of quantum spin-1/2 state vectors via complex-valued restricted Boltzmann machines in Sect. 5.1. There states are sampled from the represented probability distribution to approximate expectation values of quantum operators. This suggests that the sampling can be implemented on the neuromorphic chips, from which we expect an immense speedup. However, since the restricted Boltzmann machine used for the parametrization consists of complex weights and biases, the spin states do not follow a Boltzmann distribution and can thus not straightforwardly be sampled on the hardware. Hence, to enable an implementation we need to adapt the state sampling ansatz and recover a Boltzmann distribution for the visible and hidden neurons in the complex-valued restricted Boltzmann machine. Additionally, we need to find a way to include the complex phases.

We suggest such an adaptation, the phase-reweighted sampling scheme [5–8], in this chapter and discuss its benefits and limitations. Furthermore, we introduce deep neural networks to perform measurements of operators in any cartesian basis by only evaluating visible variables. We then benchmark the ansatz on ground states in the transverse-field Ising model, as well as on Bell and Greenberger-Horne-Zeilinger states.

This chapter contains discussions and results based on [9].

S. Czischek, *Neural-Network Simulation of Strongly Correlated Quantum Systems*,
Springer Theses, https://doi.org/10.1007/978-3-030-52715-0_6

## 6.1   The Phase Reweighting Ansatz

In Sect. 5.1 we have introduced the state vector parametrization of quantum spin-1/2 systems via a complex-valued restricted Boltzmann machine (RBM). We are now interested in implementing the sampling of spin configurations from the represented probability distribution on the neuromorphic hardware of the BrainScaleS group, as introduced in Sect. 3.5. From this we expect a speedup in creating samples to approximately evaluate expectation values of general quantum operators. Since the hardware can efficiently draw samples from Boltzmann distributions, we propose a sampling scheme for the complex-valued RBM which absorbs the complex phases into the sampled observables. Thus, it defines a Boltzmann distribution underlying the visible and hidden variables. This so-called phase reweighting scheme is commonly used in quantum Monte Carlo methods [5–8] and enables the application of block Gibbs sampling in the complex-valued RBM, as introduced in Sect. 3.2.2. Furthermore, it allows an extension to deep networks, since it defines a probability distribution for the hidden variables. This was the limiting factor for such an extension in the complex-valued RBM as discussed in Sect. 5.1.1. We make use of this possibility in Sect. 6.2 where we directly parametrize spin states in any cartesian basis using deep neural networks.

Despite these advantages the phase reweighting scheme is known to experience the sign problem, which leads to an exponentially scaling number of samples necessary to reach a certain accuracy and is caused by highly fluctuating phase factors. This makes the sampling computationally expensive and limits the ansatz to small system sizes [5–8]. On the other hand, an implementation of the sampling on neuromorphic hardware can still provide a proportional speedup compared to classical computers.

To derive the phase reweighting scheme, we take a closer look at the complex-valued RBM introduced in Sect. 5.1. The exponential of the network energy provides the complex coefficients,

$$p(\mathbf{v}, \mathbf{h}; \mathcal{W}) = \exp\left[\sum_{i=1}^{N}\sum_{j=1}^{M} v_i W_{i,j} h_j + \sum_{i=1}^{N} v_i d_i + \sum_{j=1}^{M} h_j b_j\right]. \qquad (6.1)$$

These can be expressed as the product of a term depending on the real parts of the weights and a term depending on the imaginary parts of the weights,

$$p(\mathbf{v}, \mathbf{h}; \mathcal{W}) = \exp\left[\sum_{i=1}^{N}\sum_{j=1}^{M} v_i W_{i,j}^{\mathrm{R}} h_j + \sum_{i=1}^{N} v_i d_i^{\mathrm{R}} + \sum_{j=1}^{M} h_j b_j^{\mathrm{R}}\right]$$

$$\times \exp\left[i\left(\sum_{i=1}^{N}\sum_{j=1}^{M} v_i W_{i,j}^{\mathrm{I}} h_j + \sum_{i=1}^{N} v_i d_i^{\mathrm{I}} + \sum_{j=1}^{M} h_j b_j^{\mathrm{I}}\right)\right] \qquad (6.2)$$

$$=: \tilde{P}\left(\mathbf{v}, \mathbf{h}; \mathcal{W}^{\mathrm{R}}\right) e^{i\tilde{\varphi}(\mathbf{v}, \mathbf{h}; \mathcal{W}^{\mathrm{I}})},$$

with $\mathcal{W} = \mathcal{W}^R + i\mathcal{W}^I$. Here we introduce the phase $\tilde{\varphi}(\mathbf{v}, \mathbf{h}; \mathcal{W}^I)$ and a real non-negative function $\tilde{P}(\mathbf{v}, \mathbf{h}; \mathcal{W}^R)$. The latter defines a probability distribution over the visible and hidden neurons, since it takes the form of the unnormalized probability distribution defined by a real-valued RBM.

Considering the RBM parametrization of a wave function as introduced in Sect. 5.1, the complex basis-expansion coefficients $c_{\mathbf{v}}(\mathcal{W})$ can be expressed as

$$c_{\mathbf{v}}(\mathcal{W}) = \sum_{\{\mathbf{h}\}} \tilde{P}\left(\mathbf{v}, \mathbf{h}; \mathcal{W}^R\right) e^{i\tilde{\varphi}(\mathbf{v},\mathbf{h};\mathcal{W}^I)}. \tag{6.3}$$

Thus, they are a sum over a product of an unnormalized probability distribution and a complex phase factor. Expectation values of operators $O^{\text{diag}}$, which are diagonal in the considered basis, can be calculated via

$$
\begin{aligned}
\langle O^{\text{diag}} \rangle &= \frac{1}{Z(\mathcal{W})} \sum_{\{\mathbf{v}\}} O^{\text{diag}}(\mathbf{v}) \, c_{\mathbf{v}}(\mathcal{W}) \, c_{\mathbf{v}}^*(\mathcal{W}) \\
&= \frac{1}{Z(\mathcal{W})} \sum_{\{\mathbf{v}\}} \sum_{\{\mathbf{h}\}} \sum_{\{\tilde{\mathbf{h}}\}} \left[ O^{\text{diag}}(\mathbf{v}) \, e^{i\tilde{\varphi}(\mathbf{v},\mathbf{h};\mathcal{W}^I) - i\tilde{\varphi}(\mathbf{v},\tilde{\mathbf{h}};\mathcal{W}^I)} \right] \\
&\qquad\qquad\qquad\qquad \times \tilde{P}\left(\mathbf{v}, \mathbf{h}; \mathcal{W}^R\right) \tilde{P}\left(\mathbf{v}, \tilde{\mathbf{h}}; \mathcal{W}^R\right) \\
&= \frac{1}{Z(\mathcal{W})} \sum_{\{\mathbf{v}\}} \sum_{\{\mathbf{h}\}} \sum_{\{\tilde{\mathbf{h}}\}} \left[ O^{\text{diag}}(\mathbf{v}) \, e^{i\varphi(\mathbf{v},\mathbf{h},\tilde{\mathbf{h}};\mathcal{W}^I)} \right] P\left(\mathbf{v}, \mathbf{h}, \tilde{\mathbf{h}}; \mathcal{W}^R\right),
\end{aligned}
\tag{6.4}
$$

$$Z(\mathcal{W}) = \sum_{\{\mathbf{v}\}} \sum_{\{\mathbf{h}\}} \sum_{\{\tilde{\mathbf{h}}\}} e^{i\varphi(\mathbf{v},\mathbf{h},\tilde{\mathbf{h}};\mathcal{W}^I)} P\left(\mathbf{v}, \mathbf{h}, \tilde{\mathbf{h}}; \mathcal{W}^R\right), \tag{6.5}$$

where we introduce

$$\varphi\left(\mathbf{v}, \mathbf{h}, \tilde{\mathbf{h}}; \mathcal{W}^I\right) := \tilde{\varphi}\left(\mathbf{v}, \mathbf{h}; \mathcal{W}^I\right) - \tilde{\varphi}\left(\mathbf{v}, \tilde{\mathbf{h}}; \mathcal{W}^I\right), \tag{6.6}$$

$$P\left(\mathbf{v}, \mathbf{h}, \tilde{\mathbf{h}}; \mathcal{W}^R\right) := \tilde{P}\left(\mathbf{v}, \mathbf{h}; \mathcal{W}^R\right) \tilde{P}\left(\mathbf{v}, \tilde{\mathbf{h}}; \mathcal{W}^R\right). \tag{6.7}$$

In Eq. (6.4) the division by the factor $Z(\mathcal{W})$ yields the right normalization. In both the numerator and the denominator we sum over all configurations of visible and hidden neurons. In the sum also a probability distribution underlying the states of all variables appears, so that we can approximate the sums by sampling visible and hidden configurations from $P(\mathbf{v}, \mathbf{h}, \tilde{\mathbf{h}}; \mathcal{W}^R)$,

$$\langle O^{\text{diag}} \rangle \approx \frac{\sum_{q=1}^{Q} O^{\text{diag}}\left(\mathbf{v}_q\right) e^{i\varphi\left(\mathbf{v}_q, \mathbf{h}_q, \tilde{\mathbf{h}}_q; \mathcal{W}^I\right)}}{\sum_{q=1}^{Q} e^{i\varphi\left(\mathbf{v}_q, \mathbf{h}_q, \tilde{\mathbf{h}}_q; \mathcal{W}^I\right)}}, \tag{6.8}$$

where $Q$ samples $(\mathbf{v}_q, \mathbf{h}_q, \tilde{\mathbf{h}}_q)$ are drawn. To approximate expectation values of the diagonal operator hence not only the observable itself is evaluated at the sampled configurations, but also the phase $\varphi(\mathbf{v}, \mathbf{h}, \tilde{\mathbf{h}}; \mathcal{W}^{\mathrm{I}})$. In the end the product of the evaluated observable and the complex phase factor is averaged. For the normalization we divide by the average of the evaluated phase factors, where we need to keep in mind that these can also take negative values, so that they can cancel each other. In the limit of large sample sizes they converge to the correct normalization factor.

As the observable is thus reweighted by a phase factor, this ansatz is referred to as the phase reweighting scheme, which is a standard approach in quantum Monte Carlo methods [5–8]. However, for strongly fluctuating phase factors the method is known to suffer from the sign problem due to cancellations appearing in the sum over the factors. This leads to an uncontrolled growth of the variance, which we study further in Sect. 6.4.2 [5–8].

To create samples from the distribution $P(\mathbf{v}, \mathbf{h}, \tilde{\mathbf{h}}; \mathcal{W}^{\mathrm{R}})$, a Gibbs sampling scheme as introduced in Sect. 3.2.2 can be used [10], since the probability distribution is represented by a real-valued RBM. In Eq. (6.4) we have to sum over the hidden variables twice, denoted by the sum over $\mathbf{h}$ and $\tilde{\mathbf{h}}$, with one of the sets having complex conjugate weights. This corresponds to the evaluation of the bra- and the ket-states, respectively, and can be interpreted as the neural network consisting of twice as many hidden neurons which need to be sampled individually.

With this we have introduced an approach to deal with the complex weights, where general RBM methods, such as Gibbs sampling, can be applied [10]. Thus, we get the classical network structure with Boltzmann distributions for the variables, which can hence be sampled by leaky integrate-and-fire (LIF) neurons in the neuromorphic hardware [1–4]. Additionally, this enables a generalization to deeper networks with more hidden layers.

There exist also works on representing the complex phase via an additional hidden layer in the RBM [11, 12]. As this changes the network structure, we do not consider those here but stick to the explicit evaluation of the phases.

## 6.2 Measurements in Different Bases

So far, we have introduced a way to approximately evaluate diagonal operators by sampling from a Boltzmann distribution via the phase reweighting scheme. We now derive a method to evaluate non-diagonal operators by only observing visible variables in a complex-valued RBM parametrization, without the use of local operators as discussed in Sect. 5.1.3. This overcomes the limitation to sparse operators and the necessity of explicitly evaluating the basis-expansion coefficients. The ansatz uses a deep complex network to parametrize spin states in arbitrary cartesian bases in the visible variables. Combining it with the phase reweighting approach introduced in Sect. 6.1, the sampling of the basis states can be implemented on the neuromorphic hardware, since only the states of the visible variables and the corresponding phases need to be evaluated.

Considering a general Hermitian operator $O$, we can decompose it into a string of Pauli matrices $\sigma_i^{\alpha_i}$,

$$O = \sum_{\{\alpha\}} D_\alpha \sigma_1^{\alpha_1} \otimes \cdots \otimes \sigma_N^{\alpha_N}, \qquad (6.9)$$

with $\alpha_i \in \{0, 1, 2, 3\}$ and $\sigma_i^{\alpha_i}$ acting on spin $i$, using $\sigma_i^1 = \sigma_i^x$, $\sigma_i^2 = \sigma_i^y$, $\sigma_i^3 = \sigma_i^z$ and $\sigma_i^0 = \mathbb{1}$. $D_\alpha$ are expansion coefficients. Considering a single spin-1/2 particle, we express it in the $z$-basis, such that the operator $\sigma_i^z$ acting on the single spin at site $i$ is diagonal, just as the identity operator $\mathbb{1}$. Thus, any operator which is a combination of only $\sigma_i^z$ and the identity is diagonal in the $z$-basis.

To perform a measurement of the operator $\sigma_i^x$ acting on a single particle, which is a non-diagonal operator in the $z$-basis, we can locally rotate the spin such that it is expressed in the $x$-basis and $\sigma_i^x$ becomes a diagonal operator. This can be done for each spin individually via a basis transformation,

$$\begin{aligned} |v_i^x\rangle &= \sum_{\{v_i^z\}} |v_i^z\rangle \langle v_i^z | v_i^x \rangle \\ &= \sum_{\{v_i^z\}} u_{z \to x}\left(v_i^x, v_i^z\right) |v_i^z\rangle, \end{aligned} \qquad (6.10)$$

with $|v_i^\alpha\rangle$ denoting the basis state of a single particle at site $i$ in the $\alpha$-basis, $\alpha \in \{x, y, z\}$. We introduce the rotation matrix,

$$U_{z \to x} = \frac{1}{\sqrt{2}} \begin{bmatrix} 1 & 1 \\ 1 & -1 \end{bmatrix}, \qquad (6.11)$$

with entries $u_{z \to x}(v_i^x, v_i^z) = \langle v_i^z | v_i^x \rangle$. An analogous expression is given for a rotation into the $y$-basis,

$$|v_i^y\rangle = \sum_{\{v_i^z\}} u_{z \to y}\left(v_i^y, v_i^z\right) |v_i^z\rangle, \qquad (6.12)$$

with

$$U_{z \to y} = \frac{1}{\sqrt{2}} \begin{bmatrix} 1 & i \\ i & 1 \end{bmatrix}, \qquad (6.13)$$

resulting from $u_{z \to y}(v_i^y, v_i^z) = \langle v_i^z | v_i^y \rangle$.

We can thus perform a measurement of $\sigma_i^x$ or $\sigma_i^y$ acting on a single spin by rotating it into the corresponding basis and evaluating the diagonal observable $\sigma_i^z$. The single-particle basis states $|v_i^z\rangle$ in the expansion of the state vector,

$$|\Psi\rangle = \sum_{\{\mathbf{v}^z\}} c_{\mathbf{v}}^z |v_1^z\rangle \otimes \cdots \otimes |v_N^z\rangle, \qquad (6.14)$$

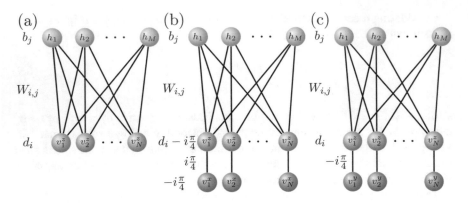

**Fig. 6.1** Visualization of the RBM and dNN setup to perform measurements in the $z$- (**a**), $x$- (**b**) and $y$-basis (**c**), where always the lowest layer contains the visible variables while all others are hidden neurons. The additional neurons $v_i^x$ and $v_i^y$ are only connected to the corresponding $v_i^z$ via an imaginary weight and for the extension into the $x$-basis additional biases appear. Figure partly adapted from [9]

are then replaced by Eq. (6.10) or Eq. (6.12), respectively, introducing a sum over all configurations of $v_i^z$. The entries of the two rotation matrices can be expressed as exponential functions,

$$u_{z \to x}\left(v_i^x, v_i^z\right) = \frac{1}{\sqrt{2}}\exp\left[i\frac{\pi}{4}\left(v_i^x v_i^z - v_i^x - v_i^z + 1\right)\right], \tag{6.15}$$

$$u_{z \to y}\left(v_i^y, v_i^z\right) = \frac{1}{\sqrt{2}}\exp\left[i\frac{\pi}{4}\left(1 - v_i^y v_i^z\right)\right]. \tag{6.16}$$

It can straightforwardly be checked that this yields the entries of the rotation matrices $U_{z \to x}$ and $U_{z \to y}$.

These exponential functions take the form of network energies as defined for an RBM. The products $i\pi/4 v^x v^z$ and $i\pi/4 v^y v^z$ indicate connecting weights of strength $i\pi/4$ between two neurons and the terms $i\pi/4 v^x$ and $i\pi/4 v^z$ correspond to biases. We thus introduce visible neurons which correspond to the spin states in different bases, so that we denote them with an additional index $\alpha \in \{x, y, z\}$. The overall bias of $i\pi/4$ in Eq. (6.15) as well as the overall factor $1/\sqrt{2}$ in front of Eqs. (6.15)–(6.16) can be neglected as they vanish in the normalization. The summation over all states of $v_i^z$ appearing in Eqs. (6.10) and (6.12) then corresponds to a marginalization over the hidden variables in the RBM, so that $v_i^z$ turns into a hidden neuron. This yields a deep neural network (dNN) structure with two hidden layers to represent spin states in the $x$- or $y$-basis. The ansatz can be generalized to spin systems with $N$ sites, where the variables $v_i^x$, $v_i^y$ can be added to any spin which is measured in the corresponding basis.

The full network takes the form as illustrated in Fig. 6.1, where Fig. 6.1a shows the standard setup with visible variables corresponding to spin states in the $z$-basis.

(b) shows the dNN extension with visible variables corresponding to spin states in the $x$-basis and (c) shows the setup with visible variables corresponding to spin states in the $y$-basis. The neurons $v_i^x$ (or analogously $v_i^y$) are added for each spin separately in a way such that the desired measured operator is represented in a diagonal basis.

Hence, a new network needs to be constructed for measurements of any non-commuting operators. This is analogous to experimental measurements, since a new sampling procedure (a new measurement) needs to be performed on a different network structure (a different experimental setup) for any operator that is measured and does not commute with a previously measured operator.

Using the phase reweighting ansatz, expectation values of the diagonal operators can be calculated by sampling from the Boltzmann distribution represented by the real parts of the weights and biases in the dNN. The observables are then evaluated at the samples and reweighted with the corresponding phase factors given in terms of the imaginary parts of the weights and biases. The expectation value of $\sigma_i^x$ acting on spin $i$ is then as an example given by

$$\langle \sigma_i^x \rangle = \frac{1}{Z(\mathcal{W})} \sum_{\{v^x\}} \sum_{\substack{\{v^z\}, \{h\}, \\ \{\tilde{v}^z\} \{\tilde{h}\}}} \left[ v_i^x e^{i\varphi\left(v^x, v^z, \tilde{v}^z, h, \tilde{h}; \mathcal{W}^I\right)} \right] P\left(v^x, v^z, \tilde{v}^z, h, \tilde{h}; \mathcal{W}^R\right),$$

(6.17)

$$Z(\mathcal{W}) = \sum_{\{v^x\}} \sum_{\substack{\{v^z\}, \{h\}, \\ \{\tilde{v}^z\} \{\tilde{h}\}}} e^{i\varphi\left(v^x, v^z, \tilde{v}^z, h, \tilde{h}; \mathcal{W}^I\right)} P\left(v^x, v^z, \tilde{v}^z, h, \tilde{h}; \mathcal{W}^R\right),$$

(6.18)

with unnormalized probability $P(v^x, v^z, \tilde{v}^z, h, \tilde{h})$ and phase $\varphi(v^x, v^z, \tilde{v}^z, h, \tilde{h})$. These are obtained from the complex dNN similarly to the RBM discussed in Sect. 6.1. As the $v_i^z$-variables are now hidden neurons, they need to be summed over twice, according to Eq. (6.4). Analogous expressions are true for measurements in the $y$-basis.

We have thus introduced a general dNN setup to perform measurements of any desired Pauli string operator, where the phase reweighting scheme can be applied to create samples from the network. With this, standard RBM methods, as introduced in Chap. 3, can be used and the sampling of the network can in principle be implemented on the neuromorphic hardware.

## 6.3 Representation of TFIM Ground States

As a first benchmark of the dNN ansatz in combination with the phase reweighting scheme we consider the transverse-field Ising model (TFIM). We stick to ground states since we are interested in the performance of the sampling scheme itself and not in the representational power or performance of the RBM parametrization. We know from Sect. 5.2.2 that TFIM ground states can be represented with good accuracy.

We consider the TFIM Hamiltonian,

$$H_{\text{TFIM}} = -\sum_{i=1}^{N} \sigma_i^z \sigma_{(i+1)\bmod N}^z - h_t \sum_{i=1}^{N} \sigma_i^x, \tag{6.19}$$

with periodic boundary conditions, and parametrize the ground state at the quantum critical point, $h_t = h_{t,c} = 1$. For this transverse field the entanglement entropy grows logarithmically with the system size and reaches its maximum [13]. This way we can check whether quantum effects are represented in the dNN ansatz. To find the representation of this state in the RBM parametrization, we set up a network with $N$ visible and $M = N$ hidden variables. We perform the training of the weights by minimizing the system energy via stochastic reconfiguration, as discussed in Sect. 5.1.2. Since the Hamiltonian of the TFIM is stoquastic, meaning that it only has non-positive off-diagonal entries, its ground states can be expressed with purely real basis-expansion coefficients [14, 15]. In the following we choose a real-valued RBM to parametrize the TFIM ground state and only the weights and biases coming up when going to a dNN to perform measurements in the $x$- or $y$-basis take imaginary values.

We consider moderate system sizes, $N \leq 10$, so that the training of the RBM can be done without sampling, but with explicitly summing over all possible spin configurations. This provides a more accurate representation of the ground state, see Sect. 5.2.2. We then fix the learned weights and biases and perform the phase reweighting scheme with block Gibbs sampling in a dNN to approximate expectation values of magnetizations and nearest-neighbor correlations in the $x$- and $z$-basis.

When benchmarking these software simulations, two sources of imperfections need to be distinguished. On the one hand, as seen in Sect. 5.2.2, the representation of the ground state in the RBM is not exact and leads to a representation error. On the other hand, statistical errors due to finite sample sizes appear in the evaluation of operators, causing sampling errors. We analyze both errors individually in the following, focusing on the sampling error.

The results are shown in Fig. 6.2, where we vary the system size from $N = 2$ to $N = 10$ in steps of one. Panel (a) analyzes the sampling error and shows the absolute deviations between sampling from the dNN with phase reweighting ($\langle O \rangle_{\text{dNN}}$) and summing over all configurations explicitly in the RBM parametrization ($\langle O \rangle_{\text{sum}}$). The deviations are plotted as functions of the sample size for magnetizations and nearest-neighbor correlations in the $x$- and $z$-basis. When summing over all spin states, operators in the $x$-basis are evaluated via the local operator approach, as discussed in Sect. 5.1.3. Here we only consider the magnetizations of the first spin and the correlations between the first two spins since we explicitly include translation invariance in the RBM parametrization, as discussed in Sect. 5.1.4.

The sampling is performed ten times for each system size and the absolute deviations are averaged, with shaded regions denoting statistical fluctuations. We find

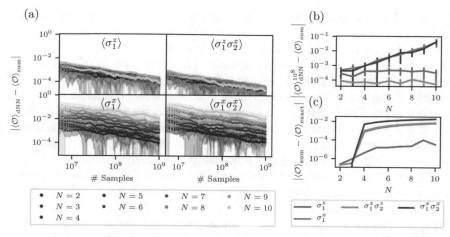

**Fig. 6.2** RBM representation of the ground state in the TFIM at the quantum critical point, $h_{t,c} = 1$, using the dNN ansatz with phase reweighting scheme, where Gibbs sampling is applied. Panel **a** analyzes the sampling error and shows absolute deviations of simulations when evaluating the RBM parametrization via sampling ($\langle O \rangle_{dNN}$) or via explicitly summing over all possible configurations ($\langle O \rangle_{sum}$). Expectation values of magnetizations of the first spin and of the correlations between the first two spins in the $x$- and $z$-basis are considered, and deviations are plotted as functions of the sample size. The chain length is varied from $N = 2$ to $N = 10$ sites, indicated by the different colors, and for each system size the sampling is performed ten times and averaged, with shaded regions denoting statistical fluctuations. For measurements in the $z$-basis (upper row) approximately no dependence on the system size is observed, while the deviations increase with the chain length for measurements in the $x$-basis (lower row). Panel **b** shows the absolute deviation of the same simulation data evaluated at $10^8$ samples ($\langle O \rangle_{dNN}^{10^8}$) from the solution when summing over all configurations as a function of system size, with error bars again denoting statistical fluctuations. An exponential scaling of the deviations with the system size is clearly visible for measurements in the $x$-basis. Panel **c** analyzes the representation error of the RBM parametrization by showing the absolute deviation when explicitly summing over all configurations in the RBM parametrization ($\langle O \rangle_{sum}$) and the exact result calculated via exact diagonalization ($\langle O \rangle_{exact}$). The representational error is found to dominate the sampling error for $N \geq 4$. Figure adapted from [9]

convergence of the absolute deviations towards zero proportional to $1/\sqrt{\text{\# Samples}}$ for measurements in both bases, as it is expected from statistical arguments [16]. When considering measurements in the $z$-basis (upper row), where the weights and biases in the RBM are purely real, we do observe approximately no dependence of the absolute deviation on the system size. When considering measurements in the $x$-basis (lower row) where imaginary weights and biases appear in the dNN, larger deviations are found. This is reasonable, since the complex phase factors can cancel each other in the normalization factor $Z(\mathcal{W})$, see Eq. (6.4). Large fluctuations can then lead to the sign problem, which causes an exponential scaling of the amount of samples necessary to reach a certain accuracy with the system size. The absolute deviation is indeed found to grow with the system size in the lower row of Fig. 6.2a, but a convergence to the exact solution is still observed for the system sizes considered.

To analyze this dependency further, panel (b) shows the same data evaluated at $10^8$ samples (indicated by the upper index $\langle O \rangle_{\mathrm{dNN}}^{10^8}$) as a function of the system size. As expected, the measurements in the $z$-basis approximately follow a constant line, while the absolute deviations of measurements in the $x$-basis scale exponentially with the system size. This renders the ansatz inefficient, even though the number of necessary samples to reach good accuracy is smaller than the number of possible configurations. This is here $2^{5N}$, since we choose $M = N$ and we have two hidden layers which are summed over twice according to Eq. (6.4). Thus, the ansatz is limited to small system sizes, but it is possible to simulate spin chains slightly longer than what is possible with exact diagonalization using comparable resources.

Figure 6.2c shows the absolute deviation of the correlations and magnetizations in the $x$- and $z$-basis when explicitly summing over all states in the RBM parametrization ($\langle O \rangle_{\mathrm{sum}}$) from the exact expectation values calculated via exact diagonalization ($\langle O \rangle_{\mathrm{exact}}$). This corresponds to the representation error and we find a sudden increase in the deviation when going from $N = 3$ to $N = 4$ sites for all observables. The deviations then saturate around $10^{-2}$ for larger system sizes, still showing good accuracy. The sudden increase results from the representational power of the RBM parametrization, so up to $N = 3$ the TFIM ground state can be represented with good accuracy and for larger $N$ it is harder to represent this state. The behavior for larger system sizes shows that it is not necessary to draw as many samples as considered in the phase-reweighted simulations, since we cannot exceed the representational accuracy.

In summary we show that the ground state of the TFIM at the quantum critical point can be represented by the dNN ansatz in combination with sampling via the phase reweighting scheme and expectation values of observables in the $x$- and $z$-basis can be approximated with good accuracy. However, the necessary number of samples to reach a certain accuracy scales exponentially with the system size for measurements in the $x$-basis, rendering the ansatz inefficient and limiting it to small system sizes.

## 6.4 Sampling a Bell State

### 6.4.1 Representing a Bell State with an RBM

Having benchmarked the dNN ansatz with phase reweighting on the ground state of the TFIM at the quantum critical point, we are now interested in the performance of the ansatz for strongly entangled states. Therefore, we consider the Bell state as introduced in Sect. 2.4.1,

$$|\Psi^{\mathrm{BP}}\rangle = \frac{1}{\sqrt{2}} \left( |\uparrow\downarrow\rangle + |\downarrow\uparrow\rangle \right). \tag{6.20}$$

To represent this state with an RBM, we can construct a network with two visible variables corresponding to the two spin particles. We additionally show in the following that it is sufficient to use one hidden variable. We again focus on analyzing whether sampling the deep network with the phase reweighting scheme works well and efficiently. Therefore, we do not variationally search for the weights to parametrize the Bell state, but we calculate them analytically and fix them.

The ansatz for finding the weights parametrizing a Bell state is given by the state vector and its RBM representation,

$$
\begin{aligned}
|\Psi^{\mathrm{BP}}\rangle &= \sum_{\{\mathbf{v}^z\}} c_{\mathbf{v}^z}(\mathcal{W}) |\mathbf{v}^z\rangle \\
&= \frac{1}{\sqrt{2}} [|-1, 1\rangle + |1, -1\rangle],
\end{aligned}
\tag{6.21}
$$

where we encode the up- and down-state of the spins in binary values with $|\uparrow\rangle = |1\rangle$, $|\downarrow\rangle = |-1\rangle$. This yields

$$
c_{\mathbf{v}^z=(\pm 1, \pm 1)}(\mathcal{W}) \overset{!}{=} 0,
\tag{6.22}
$$

$$
c_{\mathbf{v}^z=(\pm 1, \mp 1)}(\mathcal{W}) \overset{!}{=} \frac{1}{\sqrt{2}}.
\tag{6.23}
$$

Using the definition of $c_{\mathbf{v}^z}(\mathcal{W})$ according to Eq. (5.6) gives

$$
\begin{aligned}
c_{\mathbf{v}^z}(\mathcal{W}) &= \sum_{\{\mathbf{h}\}} \exp\left[\sum_{i=1}^{N}\sum_{j=1}^{M} v_i^z W_{i,j} h_j + \sum_{i=1}^{N} d_i v_i^z + \sum_{j=1}^{M} b_j h_j\right] \\
&= \exp\left[\sum_{i=1}^{N} d_i v_i^z\right] \prod_{j=1}^{M} 2\cosh\left[\sum_{i=1}^{N} v_i^z W_{i,j} + b_j\right],
\end{aligned}
\tag{6.24}
$$

which yields

$$
\exp\left[\pm (d_1 + d_2)\right] 2\cosh\left[\pm (W_{1,1} + W_{2,1}) + b_1\right] \overset{!}{=} 0,
\tag{6.25}
$$

$$
\exp\left[\pm (d_1 - d_2)\right] 2\cosh\left[\pm (W_{1,1} - W_{2,1}) + b_1\right] \overset{!}{=} \frac{1}{\sqrt{2}}.
\tag{6.26}
$$

In the last lines we use the network structure of the Bell state with $N = 2$, $M = 1$, and plug it into Eqs. (6.22)–(6.23).

There are many possible solutions to Eqs. (6.25)–(6.26), but we are only interested in finding at least one of them. Therefore, we use Eq. (6.25) to choose the ansatz

$$\cosh\left[\pm\left(W_{1,1} + W_{2,1}\right) + b_1\right] \stackrel{!}{=} 0 \tag{6.27}$$

$$\Rightarrow b_1 \pm \left(W_{1,1} + W_{2,1}\right) = i\pi\left(n_\pm + \frac{1}{2}\right), \quad n_\pm \in \mathbb{Z}. \tag{6.28}$$

This yields the expressions

$$b_1 = i\pi\left(\frac{n_+ + n_-}{2} + \frac{1}{2}\right), \tag{6.29}$$

$$W_{1,1} + W_{2,1} = i\pi\,\frac{n_+ - n_-}{2}. \tag{6.30}$$

Those can be plugged into Eq. (6.26) and by introducing $n^{(\pm)} = n_+ \pm n_-$ we get

$$\cosh\left[b_1 \pm \left(W_{1,1} - W_{2,1}\right)\right]$$
$$= \begin{cases} \pm i\,(-1)^{n^{(+)}/2}\sinh\left[W_{1,1} - W_{2,1}\right], & \text{if } n^{(\pm)} \text{ even}, \\ (-1)^{(n^{(+)}+1)/2}\cosh\left[W_{1,1} - W_{2,1}\right], & \text{if } n^{(\pm)} \text{ odd}. \end{cases} \tag{6.31}$$

We consider the two cases individually in the following.
   Using

$$\exp\left[\pm(d_1 - d_2)\right] = \cosh\left[d_1 - d_2\right] \pm \sinh\left[d_1 - d_2\right], \tag{6.32}$$

we get from Eq. (6.23) for the case of even $n^{(\pm)}$ two equations,

$$i\,(-1)^{n^{(+)}/2}\cosh\left[d_1 - d_2\right]\sinh\left[W_{1,1} - W_{2,1}\right] \stackrel{!}{=} 0, \tag{6.33}$$

$$i\,(-1)^{n^{(+)}/2}\sinh\left[d_1 - d_2\right]\sinh\left[W_{1,1} - W_{2,1}\right] \stackrel{!}{=} \frac{1}{\sqrt{8}}. \tag{6.34}$$

Thus,

$$\cosh\left[d_1 - d_2\right] \stackrel{!}{=} 0 \tag{6.35}$$

$$\Rightarrow d_1 - d_2 = i\pi\left(m_e + \frac{1}{2}\right), \quad m_e \in \mathbb{Z} \tag{6.36}$$

$$\Rightarrow (-1)^{n^{(+)}/2+1+m_e}\sinh\left[W_{1,1} - W_{2,1}\right] \stackrel{!}{=} \frac{1}{\sqrt{8}} \tag{6.37}$$

$$\Rightarrow W_{1,1} - W_{2,1} = (-1)^{n^{(+)}/2+m_e+1}\,\mathrm{arsinh}\left(\frac{1}{\sqrt{8}}\right), \tag{6.38}$$

where we use $\sinh[i\pi(m_e + 1/2)] = i(-1)^{m_e}$.
Putting this together with Eqs. (6.29)–(6.30), we obtain the result

$$d_1 - d_2 = i\pi \left( m_e + \frac{1}{2} \right), \quad m_e \in \mathbb{Z},$$

$$b_1 = i\pi \left( \frac{n_+ + n_-}{2} + \frac{1}{2} \right), \quad n_\pm \in \mathbb{Z},$$

$$W_{1,1} = (-1)^{n^{(+)}/2 + m_e + 1} \frac{1}{2} \operatorname{arsinh} \left[ \frac{1}{\sqrt{8}} \right] + i \frac{\pi}{4} n^{(-)}, \quad n^{(\pm)} = n_+ \pm n_-,$$

$$W_{2,1} = (-1)^{n^{(+)}/2 + m_e} \frac{1}{2} \operatorname{arsinh} \left[ \frac{1}{\sqrt{8}} \right] + i \frac{\pi}{4} n^{(-)}.$$

(6.39)

Analogously, we get for the case of odd $n^{(\pm)}$,

$$(-1)^{(n^{(+)}+1)/2} \cosh [d_1 - d_2] \cosh \left[ W_{1,1} - W_{2,1} \right] \overset{!}{=} \frac{1}{\sqrt{8}}, \tag{6.40}$$

$$(-1)^{(n^{(+)}+1)/2} \sinh [d_1 - d_2] \cosh \left[ W_{1,1} - W_{2,1} \right] \overset{!}{=} 0, \tag{6.41}$$

which yields

$$\sinh [d_1 - d_2] = 0 \tag{6.42}$$

$$\Rightarrow d_1 - d_2 = i\pi m_o, \quad m_o \in \mathbb{Z}. \tag{6.43}$$

Using $\cosh[i\pi m_o] = (-1)^{m_o}$, we obtain from Eq. (6.40),

$$(-1)^{(n^{(+)}+1)/2 + m_o} \cosh \left[ W_{1,1} - W_{2,1} \right] \overset{!}{=} \frac{1}{\sqrt{8}} \tag{6.44}$$

$$\Rightarrow W_{1,1} - W_{2,1} = i \left[ \pi \left( \frac{n^{(+)} + 1}{2} + m_o \right) + (-1)^{(n^{(+)}+1)/2 + m_o} \arccos \left( \frac{1}{\sqrt{8}} \right) \right]. \tag{6.45}$$

In the end this provides the result for odd $n^{(\pm)}$,

$$d_1 - d_2 = i\pi m_o, \quad m_o \in \mathbb{Z},$$

$$b_1 = i\pi \left( \frac{n_+ + n_-}{2} + \frac{1}{2} \right), \quad n_\pm \in \mathbb{Z}, \quad n^{(\pm)} = n_+ \pm n_-$$

$$W_{1,1} = i \left[ (-1)^{(n^{(+)}+1)/2 + m_o} \frac{1}{2} \arccos \left( \frac{1}{\sqrt{8}} \right) + \frac{\pi}{4} (2n_+ + 1 + 2m_o) \right],$$

$$W_{2,1} = i \left[ (-1)^{(n^{(+)}+1)/2 + m_o + 1} \frac{1}{2} \arccos \left( \frac{1}{\sqrt{8}} \right) - \frac{\pi}{4} (2n_- + 1 + 2m_o) \right].$$

(6.46)

These expressions for even and odd $n^{(\pm)}$ give infinitely many possible weight configurations to parametrize the Bell state due to periodicity in the hyperbolic cosine.

Thus, the same state can be represented with infinitely many choices, providing a degeneracy in the quantum state representation.

In the expressions for the weights one can see that they can be chosen either complex or even purely imaginary. These are two cases we consider in the following, where in the purely imaginary case all spin configurations are sampled with equal probabilities in the phase reweighting scheme.

To represent a Bell state with purely imaginary weights, we choose $n_+ = -1$, $n_- = 0$, so that $n^{(\pm)}$ is odd. Together with $m_o = 0$ this gives,

$$d_1 - d_2 = 0 \;\Rightarrow\; d_1 = d_2 = 0,$$
$$b_1 = 0,$$
$$W_{1,1} = i \left( -\frac{1}{2} \arccos \left[ \frac{1}{\sqrt{8}} \right] - \frac{\pi}{4} \right), \tag{6.47}$$
$$W_{2,1} = i \left( \frac{1}{2} \arccos \left[ \frac{1}{\sqrt{8}} \right] - \frac{\pi}{4} \right).$$

To represent the Bell state with complex weights we choose $n_+ = 1$, $n_- = -1$, so that $n^{(\pm)}$ is even. We additionally choose $m_e = 0$, so that we get

$$d_1 - d_2 = i\frac{\pi}{2} \;\Rightarrow\; d_1 = i\frac{\pi}{2}, \; d_2 = 0,$$
$$b_1 = i\frac{\pi}{2},$$
$$W_{1,1} = -\frac{1}{2} \operatorname{arsinh} \left( \frac{1}{\sqrt{8}} \right) + i\frac{\pi}{2}, \tag{6.48}$$
$$W_{2,1} = \frac{1}{2} \operatorname{arsinh} \left( \frac{1}{\sqrt{8}} \right) + i\frac{\pi}{2}.$$

These are two choices of weights we use to benchmark the phase-reweighted sampling scheme and the dNN approach in the following. Since these weights have been derived analytically, they represent the Bell state exactly and no representation error appears.

## 6.4.2  Violating Bell's Inequality

Given the network structure and the weights to represent the Bell state we can check how the sampling via phase reweighting performs. As the system consists only of two sites, it can straightforwardly be solved exactly and we can calculate the expected magnetizations and correlations in the $z$- and $x$-basis,

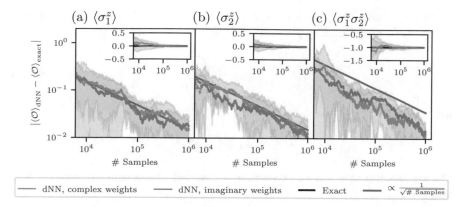

**Fig. 6.3** Expectation values resulting from the RBM representation of a Bell state using the phase reweighting scheme. We consider the magnetization $\langle \sigma_i^z \rangle$ of the first (**a**) and the second spin (**b**) and the correlation $\langle \sigma_1^z \sigma_2^z \rangle$ between the two (**c**) in the $z$-basis. Results are shown for the purely imaginary (orange) and the complex (green) choice of weights [Eqs. (6.47)–(6.48)]. Main plots show the absolute deviations of the simulations from the exact solution, while the insets show the direct measurement outcome as a function of sample size together with the exact solution. $x$-axes of the insets are the same as in the main plots and hence show the number of samples, while the $y$ axes of the insets directly correspond to the evaluated observables. Shaded regions denote statistical fluctuations from averaging ten simulation runs. The blue line in the main plot is a function proportional to $1/\sqrt{\text{# Samples}}$, indicating the expected convergence, where the proportionality factor is calculated explicitly [Eq. (6.56)]. Figure partly adapted from [9]

$$
\begin{aligned}
\langle \sigma_1^z \rangle &= \langle \sigma_2^z \rangle = 0, \\
\langle \sigma_1^z \sigma_2^z \rangle &= -1, \\
\langle \sigma_1^x \rangle &= \langle \sigma_2^x \rangle = 0, \\
\langle \sigma_1^x \sigma_2^x \rangle &= 1.
\end{aligned}
\tag{6.49}
$$

We implement the RBM with the weights as derived [Eqs. (6.47)–(6.48)] and apply Gibbs sampling to draw the visible and hidden variables from the Boltzmann distribution defined by the real parts of the weights. We then explicitly calculate the corresponding phases using the imaginary parts of the weights and apply the reweighting scheme when calculating expectation values.

The results for measurements in the $z$-basis, where we only have a shallow two-layer RBM, are shown in Fig. 6.3. Calculations for complex and purely imaginary weights are plotted for the magnetization of each spin [Fig. 6.3a, b] and their correlation [Fig. 6.3c]. The insets show the direct outcomes of the measurements together with the exact solution as a function of sample size. The main plots show the absolute deviations from the exact result together with a function proportional to $1/\sqrt{\text{# Samples}}$, indicating the expected convergence for Monte Carlo sampling [16].

As we consider small system sizes, the proportionality factor in the expected convergence behavior can be calculated explicitly. The absolute deviation of the expectation value of an observable $X$, approximated via Monte Carlo sampling, from the exact solution $X_{\text{exact}}$ is given by [16]

$$|\langle X \rangle - X_{\text{exact}}| = \frac{\sqrt{\sigma^2[X]}}{\sqrt{N}}, \qquad (6.50)$$

with variance

$$\sigma^2[X] = \langle X^2 \rangle - \langle X \rangle^2. \qquad (6.51)$$

The angular brackets denote expectation values calculated using $Q$ Monte Carlo samples $X_q$,

$$\langle X \rangle = \frac{1}{Q} \sum_{q=1}^{Q} X_q. \qquad (6.52)$$

In the case of evaluating a general diagonal operator $\langle O \rangle_{\text{dNN}}$ via averaging over samples drawn from the RBM representation, error propagation yields

$$\sigma\left[\langle O \rangle_{\text{dNN}}\right] = \left| \frac{\sigma\left[\text{Re}\langle \Psi | O | \Psi \rangle_{\text{dNN}}\right]}{\text{Re}\langle \Psi | \Psi \rangle_{\text{dNN}}} \right|$$
$$+ \left| \frac{\text{Re}\langle \Psi | O | \Psi \rangle_{\text{dNN}}}{\text{Re}\langle \Psi | \Psi \rangle_{\text{dNN}}^2} \sigma\left[\text{Re}\langle \Psi | \Psi \rangle_{\text{dNN}}\right] \right|. \qquad (6.53)$$

Here we use

$$\langle O \rangle_{\text{dNN}} = \text{Re}\left[ \frac{\langle \Psi | O | \Psi \rangle_{\text{dNN}}}{\langle \Psi | \Psi \rangle_{\text{dNN}}} \right]$$
$$= \text{Re}\left[ \frac{\sum_{\{\mathbf{v},\mathbf{h},\tilde{\mathbf{h}}\}} O(\mathbf{v})\, p(\mathbf{v}, \mathbf{h}; \mathcal{W})\, p\left(\mathbf{v}, \tilde{\mathbf{h}}; \mathcal{W}^*\right)}{\sum_{\{\mathbf{v},\mathbf{h},\tilde{\mathbf{h}}\}} p(\mathbf{v}, \mathbf{h}; \mathcal{W})\, p\left(\mathbf{v}, \tilde{\mathbf{h}}; \mathcal{W}^*\right)} \right], \qquad (6.54)$$

with the network energy as defined in Eq. (6.1) or the corresponding expression for deep networks. Here $\mathbf{v}$ denotes the basis states in an arbitrary cartesian basis and $\mathbf{h}$, $\tilde{\mathbf{h}}$ denote all hidden variables, even if they correspond to different hidden layers in deep networks.

We can calculate the variances explicitly,

$$\sigma^2 \left[ \text{Re} \left( \langle \Psi \, |O| \, \Psi \rangle_{\text{dNN}} \right) \right] = \frac{1}{2^N} \sum_{\{\mathbf{v}, \mathbf{h}, \tilde{\mathbf{h}}\}} \text{Re} \left[ O(\mathbf{v}) \, p(\mathbf{v}, \mathbf{h}; \mathcal{W}) \, p\left(\mathbf{v}, \tilde{\mathbf{h}}; \mathcal{W}^*\right) \right]^2$$

$$- \left( \frac{1}{2^N} \sum_{\{\mathbf{v}, \mathbf{h}, \tilde{\mathbf{h}}\}} \text{Re} \left[ O(\mathbf{v}) \, p(\mathbf{v}, \mathbf{h}; \mathcal{W}) \, p\left(\mathbf{v}, \tilde{\mathbf{h}}; \mathcal{W}^*\right) \right] \right)^2,$$

$$\sigma^2 \left[ \text{Re} \left( \langle \Psi \, | \, \Psi \rangle_{\text{dNN}} \right) \right] = \frac{1}{2^N} \sum_{\{\mathbf{v}, \mathbf{h}, \tilde{\mathbf{h}}\}} \text{Re} \left[ p(\mathbf{v}, \mathbf{h}; \mathcal{W}) \, p\left(\mathbf{v}, \tilde{\mathbf{h}}; \mathcal{W}^*\right) \right]^2$$

$$- \left( \frac{1}{2^N} \sum_{\{\mathbf{v}, \mathbf{h}, \tilde{\mathbf{h}}\}} \text{Re} \left[ p(\mathbf{v}, \mathbf{h}; \mathcal{W}) \, p\left(\mathbf{v}, \tilde{\mathbf{h}}; \mathcal{W}^*\right) \right] \right)^2.$$

$$(6.55)$$

Since we only evaluate operators in the bases where they are diagonal, these expressions are generally true. For small system sizes the sum over all configurations can be evaluated explicitly, so that we can calculate the variances exactly. This exact sum over all basis states also results in the imaginary parts of the expectation values to vanish with high accuracy, which is why we do not consider them in the calculations above. For large enough sample sizes, when the effects resulting from fluctuating phase factors are suppressed, we expect the absolute deviations of the simulations from the exact solution to decrease as predicted by the variances,

$$|\langle O \rangle_{\text{dNN}} - \langle O \rangle_{\text{exact}}| = \frac{\sigma \left[ \langle O \rangle_{\text{dNN}} \right]}{\sqrt{\# \text{ Samples}}}. \qquad (6.56)$$

We have evaluated the variances explicitly for the Bell state and the blue lines in Fig. 6.3 show the results. There we only plot the expected convergence calculated for the case of purely imaginary weights, as the complex weights provide similar results. A convergence to the exact solution is clearly visible and the absolute deviations decay as expected, following the blue line accurately. Increasing the sample size further leads to even smaller absolute deviations but is also computationally more expensive. The deviation decreases slowly with the number of samples according to the observed convergence behavior. No clear difference between the cases of complex and purely imaginary weights is visible. They both converge equally well, even though the variables are sampled from uniform distributions for purely imaginary weights, while a different distribution is given by the real parts of the complex weights.

After having shown that we can represent the state correctly in the $z$-basis, we also need to check the $x$-basis to be sure that we capture the full quantum state. Therefore, we use a deep network with an additional layer and measure the magnetization of

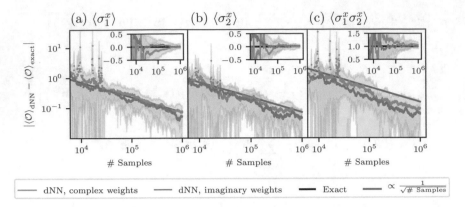

**Fig. 6.4** RBM representation in the $x$-basis using the dNN ansatz and the phase reweighting scheme to represent a Bell state. Results are shown for the case of purely imaginary (orange) and complex (green) weights [Eqs. (6.47)–(6.48)], where the main plots show the absolute deviation of the simulations from the exact solution and insets show the direct outcome of the measurements as functions of the number of samples together with the exact solution. The magnetizations $\langle \sigma_i^x \rangle$ of the first (**a**) and the second spin (**b**), as well as their correlation $\langle \sigma_1^x \sigma_2^x \rangle$ (**c**) are evaluated. $x$-axes in the insets show the sample size and are the same as in the main plots, while on the $y$-axes the direct observables are plotted. The blue line in the main plots denotes the explicitly calculated convergence according to Eq. (6.56), which is expected from statistical arguments [16]. Shaded regions show statistical fluctuations resulting from averaging ten simulation runs. Figure partly adapted from [9]

each spin and the correlation between the two in the $x$-basis. Results are shown in Fig. 6.4, which has the same structure as Fig. 6.3.

Here we can make the same observations as for the measurements in the $z$-basis, convergence is found for large sample sizes and the absolute deviation decays as expected. We again calculate the variance explicitly to get the expected behavior. However, the deviations are larger than for the measurements in the $z$-basis in Fig. 6.3. This is reasonable since the network size has increased and more neurons need to be sampled, leading to larger variances. There is still no clear difference between the two cases of complex and purely imaginary weights. Furthermore, we find fluctuations for small sample sizes in the inset plots, which are caused by the sum over the complex phase factors appearing in the denominator to get the right normalization, see Eq. (6.4). The individual terms appearing there can cancel each other so that the denominator can get very small and causes divergences. This is especially a problem when not enough samples exist to overcome such cancellations.

Having shown that correlations in the $x$- and $z$-basis are represented well by the dNN approach, we expect that also the Clauser–Horne–Shimony–Holt-inequality (CHSH-inequality) is violated. An observable violating this inequality is given by the two correlations in the case of the Bell state, as discussed in Sect. 2.4.1,

$$\mathcal{B}^{\mathrm{BP}} = \sqrt{2}\left[\langle \sigma_1^x \sigma_2^x \rangle - \langle \sigma_1^z \sigma_2^z \rangle\right]. \tag{6.57}$$

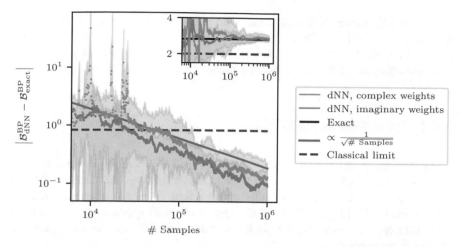

**Fig. 6.5** CHSH-observable $\mathcal{B}^{BP}$ as a function of sample size in the phase reweighting scheme applied to the RBM representation of the Bell state using purely imaginary (orange) and complex (green) weights [Eqs. (6.47)–(6.48)]. The main plot shows the absolute deviation of the simulation outcome from the exact result together with the explicitly calculated expected convergence [Eq. (6.56)], while the inset shows the CHSH-observable [Eq. (6.57)] as a function of the sample size together with the exact solution. Thus, the $x$-axis is the same as in the main plot, while the CHSH-observable is plotted on the $y$-axis. The classical limit of $\mathcal{B}^{BP}$ is shown by the red dashed line to make the violation of the CHSH-inequality visible. In the main plot this line shows the deviation of the classical limit from the exact solution, so that the CHSH-inequality is violated if the simulation goes below this line. Figure adapted from [9]

The accuracy of the simulated correlations is large enough to expect a violation.

Figure 6.5 shows the convergence of the CHSH-observable towards the exact value as a function of sample size. The main plot shows the absolute deviation from the exact solution and the inset shows the explicit measurement outcome. The classical limit is depicted by the red dashed line. In the main plot this line corresponds to the absolute deviation of the classical limit from the exact result, $2\sqrt{2} - 2$. Thus, when the absolute deviation goes below this line, the CHSH-inequality is violated, which is the case here for approximately $10^5$ and more samples. This is in agreement with observations in the inset. We hence find quantum entanglement in the state simulated via the dNN ansatz. However, explicitly summing over all configurations of the variables in the network representing the Bell state in the $z$-basis requires $2^{N+2M} = 2^4$ terms. Note that the hidden variables are summed over twice to represent the bra- and ket-states according to Eq. (6.4). Analogously, the network representing the Bell state in the $x$-basis can take $2^{N+2N+2M} = 2^8$ possible configurations in the weights. Compared to these numbers, we need many samples to violate the CHSH-inequality. This makes the ansatz computationally expensive, since for each sample the phase needs to be evaluated in the reweighting scheme. Furthermore, we need to keep in mind that we calculate the phases explicitly, so we do not get quantum entanglement directly out of the samples drawn from a classical network.

### 6.4.3  Variations in the Weights

To analyze the performance of the phase reweighting scheme in more detail, we vary the weights representing a Bell state in the RBM by some factor $\Delta \mathcal{W}$. We then check how accurately the sampled spin configurations can approximate exact expectation values of quantum operators. For such small systems we can calculate the exact expectation values by explicitly summing over all possible configurations instead of sampling them.

We first consider the purely imaginary weights parametrizing the Bell state as stated in Eq. (6.47) and set the connecting weight $W_{1,1}$ to

$$\tilde{W}_{1,1} = W_{1,1} + i \Delta W_{1,1}, \tag{6.58}$$

so that we vary it around the representation of the Bell state. Results for this are shown in Fig. 6.6, where the weight is varied in the range $\Delta W_{1,1} \in [-4, 4]$. The Bell state is represented for $\Delta W_{1,1} = 0$ and, due to the periodicity of the wave function representation in the weights, also for $\Delta W_{1,1} = \pm \pi$. Figure 6.6 shows the simulation results evaluated after $P = 10^5$ samples in comparison to the exact solution for different observables. Main panels show the absolute deviations of the simulation from the exact result and insets show the direct evaluations of the observables together with the exact solution.

Due to symmetry reasons, the exact magnetization is the same for both spins, so that we plot the results of both magnetizations in one figure, Fig. 6.6a for measurements in the $z$-basis and Fig. 6.6d for measurements in the $x$-basis. Figure 6.6b shows the correlation between the two spins measured in the $z$-basis and Fig. 6.6e accordingly shows the correlation in the $x$-basis. The combination of these two correlations provides the CHSH-observable $\mathcal{B}^{\text{BP}}$, which is shown in Fig. 6.6c. In the inset the classical limit, $\mathcal{B}^{\text{BP}} = 2$, is shown by the dashed line, so that the violation of the CHSH-inequality becomes visible.

One can observe that the accuracy reached for the measurements in the $z$-basis is higher than for measurements in the $x$-basis. This is reasonable since more variables need to be sampled in the deep network for the $x$-measurements. It also matches our previous observations. For the weights which represent the Bell state, the absolute deviation from the exact solution is larger than in other regimes of $\Delta W_{1,1}$. These larger deviations can even be seen when considering the magnetization in the $z$-basis or the correlation in the $x$-basis, which are both constant when varying the weights. Thus, it is harder to sample the entangled Bell state than other states which do not violate the CHSH-inequality with the observable considered here. However, these

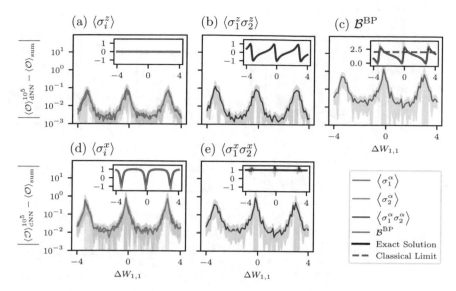

**Fig. 6.6** Simulation outcomes using the phase reweighting scheme in an RBM representing a Bell state with purely imaginary weights, where weight $W_{1,1}$ is varied by $i\Delta W_{1,1}$ around the Bell state representation [Eq. (6.47)]. Magnetizations of both spins are measured in the $z$- and $x$-basis (**a**), (**d**), respectively] besides the correlations between the two spins (**b**) $z$-basis, (**e**) $x$-basis and the CHSH-observable according to Eq. (6.57) (**c**). The main plots show the absolute deviations of the simulations from the exact solution summed over all basis states, while the insets show the direct evaluations of the observables together with the exact solution. The $x$-axes in the insets show the variation $\Delta W_{1,1}$ and are the same as in the main plots, while the direct measurement outcomes are plotted on the $y$-axes. In the inset of panel (**c**) the classical limit is shown by the dashed line to make the violation of the CHSH-inequality visible. The results are calculated using $P = 10^5$ samples, where ten sample sets are averaged. Shaded regions show the statistical fluctuations of the averages. Simulations of Bell states are found to show larger deviations than simulations of other states which do not violate the considered CHSH-inequality

can still be entangled states, as they might violate Bell's inequality for a different choice of observables.

Figure 6.7 has the same structure as Fig. 6.6 and shows the results for varying the weight $W_{1,1}$ of the analytically calculated complex weights representing the Bell state [Eq. (6.48)] according to

$$\tilde{W}_{1,1} = W_{1,1} + i\Delta W_{1,1}. \tag{6.59}$$

Due to periodicity, we again find the weights representing a Bell state for $\Delta W_{1,1} = 0$ and $\Delta W_{1,1} = \pm\pi$. In these simulations we can make the same observations as for the purely imaginary case in Fig. 6.6. We again find a larger absolute deviation from the exact solution for the weights representing the Bell state than for other choices of $\Delta W_{1,1}$.

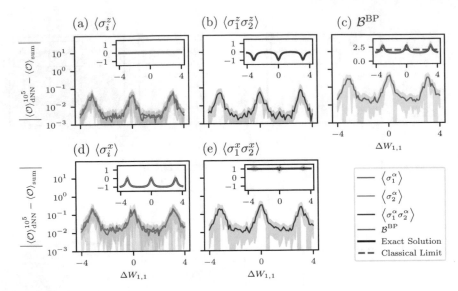

**Fig. 6.7** Expectation values of the magnetization of both spins in the $z$- (**a**) and the $x$-basis (**d**), as well as the correlation between the two spins in the $z$- (**b**) and the $x$-basis (**e**) and the CHSH-observable as stated in Eq. (6.57) (**c**) are plotted as functions of variations in the RBM weights. Simulations are made using the phase-reweighted sampling, where the evaluation is done with $P = 10^5$ samples. Ten sample sets are evaluated and averaged for each observable, shaded regions denote the statistical fluctuations. The weights are chosen complex and such that they represent a Bell state [Eq. (6.48)], where weight $W_{1,1}$ is varied by $i \Delta W_{1,1}$ in the imaginary space around the analytically calculated value. The main plots show the absolute deviations of the simulations from the exact solution summed over all basis states, while the insets show the direct evaluations of the observables together with the exact solution. In the inset of panel (**c**) the dashed line shows the classical limit of the CHSH-observable. $x$-axes in the insets are the same as in the main plots and show the variation $\Delta W_{1,1}$, while the observables are plotted on the $y$-axes

Given the complex weights, we can also make the variations of $W_{1,1}$ in the real space, namely considering

$$\tilde{W}_{1,1} = W_{1,1} + \Delta W_{1,1}. \tag{6.60}$$

Results for such simulations are shown in Fig. 6.8, which has the same structure as Figs. 6.6, 6.7. Here we do not find the periodic behavior in the weights since we vary the real part. Thus, the Bell state is only represented for $\Delta W_{1,1} = 0$. We again find a peak in the absolute deviation from the exact solution at the weights representing the Bell state, so that in the end we can conclude that it is harder to sample this state than sampling the other states we represent by varying the weights.

In Fig. 6.8 we additionally find that the absolute error increases at the corners of the plots. From this we can conclude that if the absolute value of the real part of the weights gets too large, it is also hard to sample the states. This relation to the size of the real weights is confirmed by the fact that the variations are stronger in the

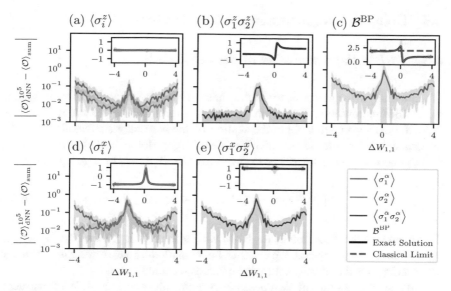

**Fig. 6.8** Real variation of the weight $W_{1,1}$ in the RBM representation by $\Delta W_{1,1}$ around the Bell state with complex weights [Eq. (6.48)] using the dNN approach. Expectation values of the magnetizations of both spins in the $z$- (**a**) and $x$-basis (**d**), of the correlations between the two spins (**b**) $z$-basis, (**e**) $x$-basis and of the CHSH-observable according to Eq. (6.57) (**c**) are calculated using the phase reweighting approach and plotted as functions of the weight variations. For calculating the expectation values, ten sample sets with $P = 10^5$ samples each are created and the measurement outcomes are averaged. Shaded regions denote statistical fluctuations. Main plots show the absolute deviations of the simulations from the exact solution summed over all basis states, while the insets show the direct measurement outcomes together with the exact solution. The $x$-axes in the insets show the variations $\Delta W_{1,1}$ and are the same as those of the main plots, while on the $y$-axes the corresponding observables are plotted. In the inset of panel (**c**) the dashed line denotes the classical limit of the CHSH-observable to make the violation of the CHSH-inequality visible. Deviations are found to increase towards the edges of the plots, where the weights take large absolute values, leading to larger numerical uncertainties

magnetization of the first spin, $\langle \sigma_1^\alpha \rangle$, whose connecting weight is varied. The effect is probably caused by the unnormalized probabilities reaching large values in these regimes, as the weights appear in an exponential function. So, for the Gibbs sampling we need to consider the ratio of two large values to get a normalized probability, which leads to larger numerical uncertainties than for the case of considering the ratio of two smaller values.

We can conclude from the variations in the weights that we should not choose the absolute values of the real parts of the weights too large to keep numerical uncertainties small. Additionally, the phase reweighting scheme shows lower accuracy for the Bell state, where Bell's inequality is maximally violated, than for states where the considered CHSH-inequality is not violated.

## 6.5 Going to Larger Systems: The GHZ State

### 6.5.1 RBM Representation of the GHZ State

We now analyze the performance of the phase-reweighted sampling scheme applied to an RBM representing larger spin systems in strongly entangled states. This provides further information about the scaling of the necessary sample size with the system size. Therefore we consider the Greenberger–Horne–Zeilinger (GHZ) state as introduced in Sect. 2.4.2 as a generalization of the Bell state to larger spin systems,

$$|\Psi^{\mathrm{GHZ}}\rangle = \frac{1}{\sqrt{2}} [|\uparrow \ldots \uparrow\rangle + |\downarrow \ldots \downarrow\rangle]. \tag{6.61}$$

To parametrize the GHZ state in the RBM we derive the weights analytically and no representation error appears. We can then benchmark the phase reweighting scheme by checking whether it can capture the exact expectation values.

We first consider the smallest possible system with $N = 3$ spins. We construct a neural network with $M = 2$ hidden variables and consider its state vector representation according to Eq. (5.6),

$$|\Psi\rangle = \sum_{\{\mathbf{v}^z\}} c_{\mathbf{v}^z}(\mathcal{W}) |\mathbf{v}^z\rangle$$

$$= \sum_{\{\mathbf{v}^z\}} \exp\left[\sum_{i=1}^{N} d_i v_i^z\right] \prod_{j=1}^{M} 2\cosh\left[\sum_{i=1}^{N} W_{i,j} v_i^z + b_j\right] |\mathbf{v}^z\rangle \tag{6.62}$$

$$\Rightarrow c_{\mathbf{v}^z=(\pm 1,\pm 1,\pm 1)}(\mathcal{W}) \overset{!}{=} \frac{1}{\sqrt{2}}, \tag{6.63}$$

$$c_{\mathbf{v}^z\neq(\pm 1,\pm 1,\pm 1)}(\mathcal{W}) \overset{!}{=} 0.$$

This provides the set of equations

$$\exp\left[\pm(d_1+d_2+d_3)\right]4\cosh\left[\pm\left(W_{1,1}+W_{2,1}+W_{3,1}\right)+b_1\right]$$
$$\times \cosh\left[\pm\left(W_{1,2}+W_{2,2}+W_{3,2}\right)+b_2\right] \overset{!}{=} \frac{1}{\sqrt{2}} \tag{6.64}$$

$$\exp\left[\pm(d_1+d_2-d_3)\right]4\cosh\left[\pm\left(W_{1,1}+W_{2,1}-W_{3,1}\right)+b_1\right]$$
$$\times \cosh\left[\pm\left(W_{1,2}+W_{2,2}-W_{3,2}\right)+b_2\right] \overset{!}{=} 0 \tag{6.65}$$

$$\exp\left[\pm(d_1-d_2+d_3)\right]4\cosh\left[\pm\left(W_{1,1}-W_{2,1}+W_{3,1}\right)+b_1\right]$$
$$\times \cosh\left[\pm\left(W_{1,2}-W_{2,2}+W_{3,2}\right)+b_2\right] \overset{!}{=} 0 \tag{6.66}$$

$$\exp\left[\pm(-d_1+d_2+d_3)\right]4\cosh\left[\pm\left(-W_{1,1}+W_{2,1}+W_{3,1}\right)+b_1\right]$$
$$\times \cosh\left[\pm\left(-W_{1,2}+W_{2,2}+W_{3,2}\right)+b_2\right] \overset{!}{=} 0 \tag{6.67}$$

There are many possibilities to solve these equations, but for now we are only interested in finding at least one possible solution. Therefore, we demand from Eq. (6.65),

$$\cosh\left[b_1 \pm \left(W_{1,1} + W_{2,1} - W_{3,1}\right)\right] \overset{!}{=} 0 \tag{6.68}$$

$$\Rightarrow b_1 = i\pi\left(\frac{n_1^+ + n_1^-}{2} + \frac{1}{2}\right), \quad n_1^{\pm} \in \mathbb{Z}, \tag{6.69}$$

$$W_{1,1} + W_{2,1} - W_{3,1} = i\pi\frac{n_1^+ - n_1^-}{2}. \tag{6.70}$$

Analogously, from Eq. (6.66) we can choose

$$\cosh\left[b_2 \pm \left(W_{1,2} - W_{2,2} + W_{3,2}\right)\right] \overset{!}{=} 0 \tag{6.71}$$

$$\Rightarrow b_2 = i\pi\left(\frac{n_2^+ + n_2^-}{2} + \frac{1}{2}\right), \quad n_2^{\pm} \in \mathbb{Z}, \tag{6.72}$$

$$W_{1,2} - W_{2,2} + W_{3,2} = i\pi\frac{n_2^+ - n_2^-}{2}, \tag{6.73}$$

and from Eq. (6.67)

$$\cosh\left[b_2 \pm \left(-W_{1,2} + W_{2,2} + W_{3,2}\right)\right] \overset{!}{=} 0 \tag{6.74}$$

$$\Rightarrow -W_{1,2} + W_{2,2} + W_{3,2} = i\pi\frac{n_2^+ - n_2^-}{2}, \tag{6.75}$$

by inserting Eq. (6.72).

This provides infinitely many solutions due to the freely choosable integers $n_1^{\pm}$ and $n_2^{\pm}$. For simplicity, and as we only need one solution, we choose $n_1^{\pm} = n_2^{\pm} = 0$ in the following. Thus,

$$\begin{aligned} b_1 = b_2 &= i\frac{\pi}{2}, \\ W_{1,1} + W_{2,1} - W_{3,1} &= 0, \\ W_{1,2} &= W_{2,2}, \\ W_{3,2} &= 0. \end{aligned} \tag{6.76}$$

This can be plugged into Eq. (6.64) to get

$$\begin{aligned} \cosh\left[i\frac{\pi}{2} \pm \left(W_{1,1} + W_{2,1} + W_{3,1}\right)\right]&\cosh\left[i\frac{\pi}{2} \pm 2W_{1,2}\right] \\ = \left(\pm i\sinh\left[W_{1,1} + W_{2,1} + W_{3,1}\right]\right)&\left(\pm i\sinh\left[2W_{1,2}\right]\right) \\ = -i\sinh\left[W_{1,1} + W_{2,1} + W_{3,1}\right]&, \end{aligned} \tag{6.77}$$

where we fix $W_{1,2} = i\pi/4$ to simplify the expression in the last line. With this, Eq. (6.64) becomes

$$-i\exp\left[\pm (d_1 + d_2 + d_3)\right]\sinh\left[W_{1,1} + W_{2,1} + W_{3,1}\right] \overset{!}{=} \frac{1}{4\sqrt{2}}. \tag{6.78}$$

Using

$$\exp\left[\pm (d_1 + d_2 + d_3)\right] = \cosh\left[d_1 + d_2 + d_3\right] \pm \sinh\left[d_1 + d_2 + d_3\right], \tag{6.79}$$

we get two equations,

$$-i\cosh\left[d_1 + d_2 + d_3\right]\sinh\left[W_{1,1} + W_{2,1} + W_{3,1}\right] \overset{!}{=} \frac{1}{4\sqrt{2}}, \tag{6.80}$$

$$-i\sinh\left[d_1 + d_2 + d_3\right]\sinh\left[W_{1,1} + W_{2,1} + W_{3,1}\right] \overset{!}{=} 0. \tag{6.81}$$

These can be solved by

$$d_1 + d_2 + d_3 = 0,$$
$$W_{1,1} + W_{2,1} + W_{3,1} = i\arcsin\left[\frac{1}{4\sqrt{2}}\right]. \tag{6.82}$$

From these constraints together with Eq. (6.70) we can extract a set of weights,

$$d_1 = d_2 = d_3 = 0,$$
$$b_1 = b_2 = i\frac{\pi}{2},$$
$$W_{1,1} = W_{2,1} = \frac{i}{4}\arcsin\left[\frac{1}{4\sqrt{2}}\right],$$
$$W_{3,1} = \frac{i}{2}\arcsin\left[\frac{1}{4\sqrt{2}}\right], \tag{6.83}$$
$$W_{1,2} = W_{2,2} = i\frac{\pi}{4},$$
$$W_{3,2} = 0.$$

From the structure of these expressions we can observe that the weights connecting to the first hidden neuron guarantee the normalization of the state vector if all spins are equal. Furthermore, they give zero coefficients if the third spin is flipped compared to the other two. On the other hand, the weights connecting to the second hidden neuron guarantee that the basis-expansion coefficients are zero if the first and second spin have opposite signs.

Having understood this structure, we can generalize it to larger spin systems and write down weights to parametrize the GHZ state as a function of the system size $N$, where we always choose $M = N - 1$ hidden variables,

$$d_l = 0,$$

$$b_j = i\frac{\pi}{2},$$

$$W_{l \neq N, 1} = \frac{i}{2(N-1)} \arcsin\left[\frac{1}{2^{N-1/2}}\right],$$

(6.84)

$$W_{N,1} = \frac{i}{2} \arcsin\left[\frac{1}{2^{N-1/2}}\right],$$

$$W_{l, j \neq 1} = i\frac{\pi}{4}\left(\delta_{l,j-1} + \delta_{l,j}\right), \quad \forall l \in \{1, \ldots, N\}, \quad j \in \{1, \ldots, M\}.$$

For small system sizes ($N \leq 5$), it has been checked numerically that these weights provide the basis-expansion coefficients of a GHZ state. Thus, we have found a set of purely imaginary weights to represent strongly entangled states of arbitrary system sizes in the RBM. We can now generally benchmark the phase reweighting scheme on those.

## 6.5.2  Sampling the GHZ State

For a first benchmark, we slightly increase the system size compared to Sect. 6.4 by considering $N = 3$ spin-1/2 particles in a GHZ state. For this we need $M = 2$ hidden variables, which need to be summed over twice with complex conjugate weights in the RBM parametrization to represent the bra- and ket-state according to Eq. (6.4). Expressing this double sum via twice as many hidden variables with complex conjugate weights, the network consists of $N + 2M = 7$ neurons when parametrizing states in the $z$-basis and of $3N + 2M = 13$ neurons when parametrizing states in the $x$-basis. This is slightly more than in the network parametrizing the Bell state, which had 4 or 8 neurons, respectively.

We set up the RBM with weights and biases as stated in Eq. (6.83) and produce samples using the phase reweighting scheme with block Gibbs sampling. With this we can measure the average magnetization per spin, the average correlation between two spins and the expectation value of operators acting on all three spins in both the $x$- and the $z$-basis. From the exact solution we expect

$$\langle \sigma_i^x \rangle = \langle \sigma_i^x \sigma_j^x \rangle = \langle \sigma_i^z \rangle = \langle \sigma_1^z \sigma_2^z \sigma_3^z \rangle = 0,$$

(6.85)

$$\langle \sigma_1^x \sigma_2^x \sigma_3^x \rangle = \langle \sigma_i^z \sigma_j^z \rangle = 1, \quad \forall i, j \in \{1, 2, 3\}, \ i \neq j.$$

(6.86)

Due to symmetries in the spin system we can average all possible combinations of spins $i$ and $j$ to calculate the expectation values, providing improved statistics in the simulations.

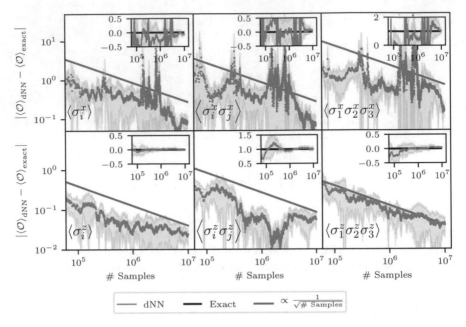

**Fig. 6.9** Measurements in the RBM parametrization of a GHZ state with $N = 3$ sites evaluated via the phase-reweighted sampling scheme in the dNN approach. Expectation values of the single-spin magnetization and of the correlation between two spins in $z$- and $x$-basis, as well as the expectation value of $\sigma^z$- and $\sigma^x$-operators acting on all three spins are shown. Due to translation invariance we average over all possible combinations of spins to perform the measurements on. The results are plotted as a function of the number of samples, where the main plots show the absolute deviation from the exact solution together with the explicitly calculated expected convergence behavior and insets show the direct evaluation of the observables on the $y$-axis with the $x$-axis being the same as in the main plots. The exact solution is shown for comparison. Five sampling chains are created and evaluated individually, where the average is considered for each observable and shaded regions show the statistical fluctuations. Figure partly adapted from [9]

Figure 6.9 shows the measurements of all six observables as functions of the sample size, where the plots have the same structure as those in Figs. 6.3, 6.4 and 6.5. One can see a clear convergence to the exact solutions for all cases in the insets and converging absolute deviations from the exact solution in the main plots. These follow the behavior expected from statistical arguments, where we can again calculate the variances explicitly, similar to the Bell state in Sect. 6.4.2 [16]. The deviations in the simulations are even slightly below the expected errors, but it can be seen that they approach the blue curve for large system sizes. For observables in the $x$-basis deviations are again larger due to the increased network size. Furthermore, large fluctuations resulting from the canceling phase factors in the denominator are visible, even though we consider $10^7$ samples. This is a huge number compared to the $2^7$ or $2^{13}$ possible network configurations for representing spin states in the $z$- or $x$-basis, respectively. We expect that even larger sample sizes would be necessary to get rid of these fluctuations and reach stable results. This shows that for the GHZ state the phase reweighting scheme is computationally expensive, as a huge amount of samples is necessary to avoid fluctuations by summing up the phase factors in the right way.

To further analyze the scaling of the necessary number of samples with the system or network size we next consider a spin chain with $N = 5$ sites in a GHZ state. Here we need $M = 4$ hidden variables, so that the neural network representing this state consists of 13 neurons for measurements in the $z$-basis and of 23 neurons for measurements in the $x$-basis. We set up the RBM with the corresponding weights and biases to represent a GHZ state according to Eq. (6.84) and use the phase reweighting scheme to sample from it. This time we consider measurements of the average magnetization per site, the average correlation between two sites and the expectation value of operators acting on all five spins in the $z$- and $x$-basis. We can again average over all spins due to symmetry reasons. For the magnetizations and the two-spin correlations the exact solutions are the same as in the case of $N = 3$ spins, while for the measurements of all five spins we expect

$$\langle \sigma_1^z \sigma_2^z \sigma_3^z \sigma_4^z \sigma_5^z \rangle = 0, \tag{6.87}$$

$$\langle \sigma_1^x \sigma_2^x \sigma_3^x \sigma_4^x \sigma_5^x \rangle = 1. \tag{6.88}$$

Figure 6.10 has the same structure as Fig. 6.9 and shows the simulation results of the five-spin GHZ state. We find the expected convergence for measurements in the $z$-basis, but the results in the $x$-basis show large fluctuations and do not converge to the exact solution. Here much more samples are expected to be necessary to get a stable result, which is in accordance with the explicitly calculated error decay. We already consider $10^{10}$ samples compared to $2^{13}$ or $2^{23}$ possible network configurations when representing spin states in the $z$- or $x$-basis, respectively. Increasing the sample size further is not possible with the computational setup we are given. This suggests an exponential scaling of the necessary sample size with the system size, rendering the phase reweighting ansatz inefficient for strongly entangled states. Due to the exponential scaling of the sample size the ansatz is limited to small systems.

## 6.6   Facing the Sign Problem

Applying the dNN approach with phase-reweighted sampling, we have found that the number of samples necessary to reach a certain accuracy scales exponentially with the system size. We have shown that this is also in accordance with the expected behavior due to statistical reasons [16]. This observation suggests the existence of a sign problem, which commonly appears in quantum Monte Carlo studies in which the problem of having complex coefficients instead of probabilities is solved by reweighting with the phase [5–8, 17, 18].

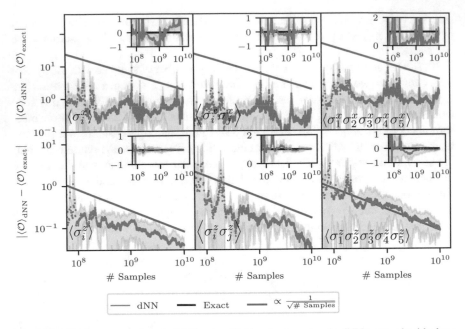

**Fig. 6.10** RBM representation of a GHZ state with $N = 5$ spins using the dNN approach with phase reweighting. Expectation values of the single-site magnetization in $z$- and $x$-basis, of the correlation between two spins in $z$- and $x$-basis and of $\sigma^z$- and $\sigma^x$-operators acting on all five spins are considered as a function of the number of samples. Five sample sets are created and averaged to calculate the individual expectation values, where the shaded regions denote the statistical fluctuations. The main plots show the absolute deviations from the exact solution together with the explicitly calculated expected convergence behavior, while the insets show the direct evaluation of the expectation values together with the exact solution. On the $y$-axes of the insets the corresponding observables are plotted, while on the $x$-axes the sample sizes are plotted, so these are the same as those of the main plots. Figure partly adapted from [9]

The sign problem comes from summing or integrating over highly fluctuating functions, so it also appears when calculating path integrals in quantum field theory. The main problem with the highly oscillating functions is that their signs also fluctuate heavily. When calculating the sum or the integral numerically, the individual terms can hence cancel or nearly cancel each other and it is hard to approximate the correct result accurately [19]. Considering the phase reweighting scheme, we sum up the phases corresponding to the samples drawn from the underlying real distribution. These phases can have different signs and thus they can also cancel each other and convergence to the exact expression requires a huge amount of samples [5–8, 17–19].

If we consider the calculation of expectation values, as stated in Eq. (6.4), we find that we also sum over the phases in the denominator to normalize the outcome. Thus, the cancellations can lead to divergent results. This shows that if the phases have absolute values close to each other but different signs, it is difficult to sum them up in the right way [5–8, 17, 18].

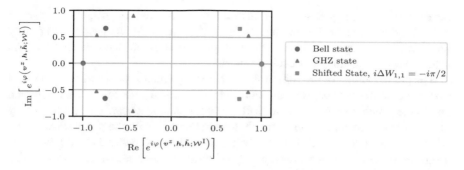

**Fig. 6.11** Scatter plot of the real and imaginary parts of the phase factors [Eq. (6.89)] belonging to the possible state configurations in an RBM representing the $z$-basis of the Bell state, the 3-spin GHZ state and a state reached via varying the weight $\tilde{W}_{1,1} = W_{1,1} + i \Delta W_{1,1}$ in the purely imaginary Bell state representation [Eq. (6.47)] by $i \Delta W_{1,1} = -i\pi/2$. The factors are distributed symmetrically with respect to the real axis and some of them are even symmetric with respect to the imaginary axis for the Bell and GHZ states. Thus, summing them up can lead to cancellations causing the sign problem. This is not the case for the shifted state, which has only positive real parts

As this is the behavior we find in the phase-reweighted simulations of spin systems, we analyze the phases in more detail. To do so we consider the individual states that can be represented by the complex-valued RBM parametrizing spin states in the $z$-basis. Figure 6.11 shows the phase factors,

$$e^{i\varphi\left(\mathbf{v}^z,\mathbf{h},\tilde{\mathbf{h}};\mathcal{W}^I\right)} = \exp\left[ i \sum_{i=1}^{N} \sum_{j=1}^{M} v_i^z W_{i,j}^I \left( h_j - \tilde{h}_j \right) + i \sum_{j=1}^{M} b_j^I \left( h_j - \tilde{h}_j \right) \right], \quad (6.89)$$

of the individual neuron configurations for a Bell state represented with purely imaginary weights and the GHZ state with $N = 3$ spins, where the weights are also purely imaginary. Furthermore, the phase factors for a state represented via varying the weight $\tilde{W}_{1,1} = W_{1,1} + i \Delta W_{1,1}$ by $i \Delta W_{1,1} = -i\pi/2$ in the purely imaginary representation of the Bell state are shown. This corresponds to a state that can be represented more accurately, as observed in Fig. 6.6. It is not visible in Fig. 6.11 which phase factor belongs to which basis state, but so far we only have a look at which factors exist. The points are scattered in the complex plane, where some states have the same phases. Thus, some of the 16 configurations for the two-spin states represented by a network with 4 neurons, or of the 128 configurations for the GHZ state represented by an RBM with 7 neurons, have phase factors lying on top of each other.

For all cases we can see that the distribution is symmetric with respect to the real axis. Hence, phase factors with same absolute values but different signs in the imaginary part exist. These factors need to be summed up in the right way to cancel the imaginary parts. Considering the imaginary axis for the Bell and GHZ states, most phase factors are also symmetric with respect to it, except for one pair of states for each of the two systems. Thus, also the sums of the real parts can get small in total and they need to be summed up in the right ratio. As the weights are purely imaginary, we sample all states with equal probabilities, so that cancellations can appear when calculating expectation values.

The phase factors for the shifted state are also symmetric with respect to the real axis, but they all have positive real values. Thus, their real parts cannot cancel, which makes the simulations more stable as divergences cannot appear. This explains why these states can be approximated much more accurately with the phase reweighting scheme. From having a look at the phase factors we hence observe that we experience a sign problem in the phase reweighting scheme as expected, which also matches our simulation results [5–8, 17, 18].

We have furthermore seen in the simulations of the TFIM ground state that the sign problem also appears in the dNN extension representing states in the $x$-basis. Hence, even for states with purely positive real basis-expansion coefficients only measurements in the $z$-basis can be evaluated efficiently. For measurements in the $x$-basis we observe a sign problem coming up.

## 6.7 Summary

In this chapter we have adapted the scheme for sampling from the complex-valued restricted Boltzmann machine (RBM) used to parametrize state vectors of quantum spin-1/2 systems. With the introduced phase-reweighted sampling scheme, the visible and hidden variables are sampled from a Boltzmann distribution. This enables the efficient evaluation of deep networks with more than one hidden layer, as well as a direct implementation of the sampling on neuromorphic hardware. In the phase reweighting scheme samples are drawn from the distribution defined by the absolute values of the complex basis-expansion coefficients. When calculating expectation values, the observables are reweighted by the corresponding phases, which need to be calculated explicitly. This is a common approach in quantum Monte Carlo methods [5–8].

For the calculation of expectation values of diagonal operators we have shown that the hidden layers need to be summed over twice with complex conjugate weights and biases. This corresponds to the representation of the bra- and ket-states in the evaluation of quantum measurements. Furthermore, we have derived a deep neural network (dNN) setup which enables evaluations of Pauli string operators in all different cartesian bases by introducing $\mathbf{v}^x$- and $\mathbf{v}^y$-layers. A separate network needs to be created for each basis a measurement is performed in. The dNN setup in combination with the phase reweighting scheme enables the applicability of standard RBM methods (Chap. 3) and also a sampling on the neuromorphic hardware (Sect. 3.5). However, we have run into new problems, conserving the amount of difficulty.

Besides such reweighting approaches being computationally expensive for large sample sizes, as the phases need to be calculated explicitly, they are also known to experience the sign problem. This is a problem resulting from heavily fluctuating phase factors, so that they have different signs and can cancel each other [5–8, 17–19]. As we also have to normalize with the sum over the phase factors in the reweighting ansatz, this can lead to divergences in the expectation values. Only in the limit of infinitely many samples, when the factors add up in the desired way, convergence to the right stable solution is guaranteed. However, the necessary number of samples to reach a certain accuracy can get huge and in the case of the sign problem it grows exponentially with system size. Thus, only small systems can be considered within suitable computation time.

When simulating small strongly entangled quantum states, such as Bell or GHZ states, with the phase reweighting scheme applied to the dNN ansatz, we have pointed out that we need exponentially many samples. This indicates the expected presence of a sign problem, which is in accordance with previous works [5–8, 17–19]. Simulating other states, such as the ground state of the transverse-field Ising model (TFIM) at the quantum critical point, we have found better convergence to the exact solution. Less fluctuations have appeared and smaller numbers of samples were necessary to reach a certain accuracy. However, even for the ground state of the TFIM we have shown that the number of necessary samples scales exponentially with the system size for measurements in the $x$-basis, rendering the ansatz inefficient.

In summary the dNN ansatz is a simple and general way to parametrize quantum states, whose representational accuracy can be improved by adding more hidden variables or even layers. Furthermore, the ansatz can be applied in higher spatial dimensions and the weights are trainable. The phase-reweighted sampling scheme can even be implemented on neuromorphic hardware. However, the ansatz is limited to small system sizes as it experiences the sign problem. From an implementation on a neuromorphic hardware we expect a shift of this limit to larger system sizes due to the efficient sampling procedure yielding a speedup. Hence, such an implementation would enable the creation of larger sample sets within suitable computation time. Even though the phases still need to be evaluated for each sample on a classical computer, limiting the speedup gained by using neuromorphic computing architectures, we expect that larger system sizes can be simulated on this hardware compared to a classical computer.

# References

1. Petrovici MA (2016) Form versus function: theory and models for neuronal substrates. Springer International Publishing, Berlin. https://doi.org/10.1007/978-3-319-39552-4
2. Petrovici MA, Bill J, Bytschok I, Schemmel J, Meier K (2016) Stochastic inference with spiking neurons in the high-conductance state. Phys Rev E 94:042312. https://link.aps.org/doi/10.1103/PhysRevE.94.042312
3. Buesing L, Bill J, Nessler B, Maass W (2011) Neural dynamics as sampling: a model for stochastic computation in recurrent networks of spiking neurons. PLoS Comput Biol 7(11):1–22. https://doi.org/10.1371/journal.pcbi.1002211
4. Kungl AF, Schmitt S, Klähn J, Müller P, Baumbach A, Dold D, Kugele A, Müller E, Koke C, Kleider M, Mauch C, Breitwieser O, Leng L, Gürtler N, Güttler M, Husmann D, Husmann K, Hartel A, Karasenko V, Grübl A, Schemmel J, Meier K, Petrovici MA (2019) Accelerated

physical emulation of bayesian inference in spiking neural networks. Front Neurosci 13:1201. https://www.frontiersin.org/article/10.3389/fnins.2019.01201

5. Troyer M, Wiese U-J (2005) Computational complexity and fundamental limitations to fermionic quantum Monte Carlo simulations. Phys Rev Lett 94:170201. https://link.aps.org/doi/10.1103/PhysRevLett.94.170201

6. Anagnostopoulos KN, Nishimura J (2002) New approach to the complex-action problem and its application to a nonperturbative study of superstring theory. Phys Rev D 66:106008. https://link.aps.org/doi/10.1103/PhysRevD.66.106008

7. Nakamura T, Hatano N, Nishimori H (1992) Reweighting method for quantum Monte Carlo simulations with the negative-sign problem. J Phys Soc Jpn 61(10):3494–3502. https://doi.org/10.1143/JPSJ.61.3494

8. Loh EY, Gubernatis JE, Scalettar RT, White SR, Scalapino DJ, Sugar RL (1990) Sign problem in the numerical simulation of many-electron systems. Phys Rev B 41:9301–9307. https://link.aps.org/doi/10.1103/PhysRevB.41.9301

9. Czischek S, Pawlowski JM, Gasenzer T, Gärttner M (2019) Sampling scheme for neuromorphic simulation of entangled quantum systems. Phys Rev B 100:195120. https://link.aps.org/doi/10.1103/PhysRevB.100.195120

10. Hinton GE (2012) A practical guide to training restricted Boltzmann machines. Springer, Berlin, pp 599–619. https://doi.org/10.1007/978-3-642-35289-8_32

11. Torlai G, Mazzola G, Carrasquilla J, Troyer M, Melko R, Carleo G (2018) Neural-network quantum state tomography. Nat Phys 14:447–450. https://doi.org/10.1038/s41567-018-0048-5

12. Torlai G, Melko RG (2018) Latent space purification via neural density operators. Phys Rev Lett 120:240503. https://link.aps.org/doi/10.1103/PhysRevLett.120.240503

13. Iglói F, Lin Y-C (2008) Finite-size scaling of the entanglement entropy of the quantum Ising chain with homogeneous, periodically modulated and random couplings. J Stat Mech: Theory Exp 2008(06):P06004. https://doi.org/10.1088%2F1742-5468%2F2008%2F06%2Fp06004

14. Kivlichan ID (2015) On the complexity of stoquastic Hamiltonians. https://pdfs.semanticscholar.org/a2e4/12ab75025df2fb7bf61d386e41c3ea72c538.pdf

15. Bravyi S, DiVencenzo DP, Oliveira R, Terhal BM (2008) The complexity of stoquastic local Hamiltonian problems. Quant Inf Comput 8(5):0361–0385. https://doi.org/10.26421/QIC8.5

16. Caflisch RE (1998) Monte Carlo and quasi-Monte Carlo methods. Acta Numer 7:1–49. https://doi.org/10.1017/S0962492900002804

17. Torlai G, Carrasquilla J, Fishman MT, Melko RG, Fisher MPA (2019) Wavefunction positivization via automatic differentiation. arXiv:1906.04654 [quant-ph]

18. Hangleiter D, Roth I, Nagaj D, Eisert J (2019) Easing the Monte Carlo sign problem. arXiv:1906.02309 [quant-ph]

19. Broecker P, Carrasquilla J, Melko RG, Trebst S (2017) Machine learning quantum phases of matter beyond the fermion sign problem. Sci Rep 7:8823. https://doi.org/10.1038/s41598-017-09098-0

# Chapter 7
# Towards Neuromorphic Sampling of Quantum States

Making use of the phase reweighting scheme to enable an implementation of sampling from a complex-valued restricted Boltzmann machine parametrizing a quantum spin-1/2 state vector on neuromorphic hardware, we have experienced the sign problem. As this problem arises from the complex coefficients in the basis-state expansion, we now consider an ansatz to circumvent the sign problem by choosing basis states such that these coefficients become purely real. The ansatz is based on positive-operator valued measures and has been discussed in [1]. There the authors show that this way a Bell state can be parametrized with a real-valued restricted Boltzmann machine. Furthermore, even mixed states can be represented with this ansatz, which was not possible with the parametrization based on a complex-valued restricted Boltzmann machine discussed in Sect. 5.1.

We discuss the application of positive-operator valued measures to parametrize state vectors with real-valued restricted Boltzmann machines in detail within this chapter and provide the way towards an implementation of sampling a Bell state representation with the neuromorphic hardware as an outlook.

## 7.1 From Complex to Real Probabilities

### 7.1.1 Positive-Operator Valued Measures

A way towards parametrizing quantum many-body systems with a real-valued restricted Boltzmann machine (RBM) is to get rid of the complex phases in the coefficients of the basis-state expansion, using so-called positive-operator valued measures (POVM). These can be chosen such that quantum states are fully described in a basis

S. Czischek, *Neural-Network Simulation of Strongly Correlated Quantum Systems*, Springer Theses, https://doi.org/10.1007/978-3-030-52715-0_7

where they only have real basis-expansion coefficients [1–3]. The POVM formalism is commonly used to perform measurements on a system if one is not interested in the state after the measurement, but only in the measurement outcome itself [2, 3]. We introduce the formalism in more detail in the following and then discuss how to use it for an RBM parametrization of spin-1/2 state vectors in the upcoming sections, before we apply it to simulate a spin system in a Bell state in Sect. 7.2.

General measurements can be described by a set of Hermitian operators $O_a$. The probability to get the outcome $a$ given the state vector $|\Psi\rangle$ of the system can be calculated via

$$P\left(a\right) = \langle \Psi \left| O_a^\dagger O_a \right| \Psi \rangle. \tag{7.1}$$

By defining $M^{(a)} = O_a^\dagger O_a$ we introduce a positive semi-definite operator $M^{(a)}$ whose expectation value gives the probability of getting the measurement outcome $a$ [2, 3].

From this we can define a set of positive semi-definite operators $M^{(a)}$, which give identity when summing over all possible measurement outcomes,

$$\sum_{\{a\}} M^{(a)} = \mathbb{1}, \tag{7.2}$$

so that they form a general quantum measurement. The probability of the system to obtain outcome $a$ can then generally be expressed using the density matrix,

$$P\left(a\right) = \text{Tr}\left[\rho M^{(a)}\right]. \tag{7.3}$$

It can be shown that the operators $M^{(a)}$ can be chosen such that $P(a)$ fully characterizes the density matrix and Eq. (7.3) is invertible [1–3].

A general $d$-dimensional quantum state is characterized by $d^2 - 1$ real parameters, so that a full description requires measurements with at least $d^2$ linearly independent outcomes to ensure normalization. Such a measurement is called informationally complete [2]. Thus, if we choose a set $\{M^{(a)}\}$ of $d^2$ projection operators, it can provide an informationally complete measurement. If we do so with the positive semi-definite operators, this provides a full description of a quantum state in the POVM representation with purely real amplitudes for each measurement outcome [1–3]. This is the basic idea of the POVM ansatz. The set $\{M^{(a)}\}$ of operators is then called the POVM, while the individual operators $M^{(a)}$ are the POVM elements. In the following we only consider cases where $M^{(a)}$ are projectors, which corresponds to a rank-1 POVM [1–3].

The POVM method is often used for quantum state tomography, since it requires a minimal number of expectation values, which reduces the redundancy and leads to faster convergence [2]. However, here we use a different application, namely we express a quantum state in terms of its POVM probabilities such that we can represent the underlying probability distribution with a real-valued RBM. Therefore, we focus on the POVM representation of a spin-1/2 system in the following [1, 2].

Considering a single spin-1/2 particle, it can either be up or down and hence its Hilbert space dimension is $d = 2$. An informationally complete POVM $\{M^{(a)}\}$ for the single particle thus needs to have $d^2 = 4$ possible measurement outcomes, which we can index with a variable having four possible values, $a \in \{0, 1, 2, 3\}$. The POVM is then generally given by [1]

$$\{M^{(a)}\} = \{M^{(0)}, M^{(1)}, M^{(2)}, M^{(3)}\}. \tag{7.4}$$

The POVM elements are factorable, so that for a system of $N$ spin-1/2 particles the total elements are given by the tensor products of the individual elements, [1]

$$M^{(a)} = M^{(a_1)} \otimes M^{(a_2)} \otimes \cdots \otimes M^{(a_N)}. \tag{7.5}$$

As the POVM is chosen informationally complete, any operator $O$ in the spin basis can be expanded as

$$O = \sum_{\{a\}} Q_O(a) M^{(a)}. \tag{7.6}$$

$Q_O(a)$ are, so far, unknown general coefficients and the sum runs over all possible configurations of single-particle POVM elements, indexed by $a \in \{0, 1, 2, 3\}^N$ [1].

If we consider the density matrix, it can be expanded as

$$\rho = \sum_{\{a\}} Q_\rho(a) M^{(a)}, \tag{7.7}$$

so that we get from Eq. (7.3),

$$
\begin{aligned}
P(a) &= \sum_{\{a'\}} Q_\rho(a') \operatorname{Tr}\left[M^{(a)} M^{(a')}\right] \\
&= \sum_{\{a'\}} Q_\rho(a') T_{a,a'}.
\end{aligned}
\tag{7.8}
$$

Here we introduce the overlap matrix $T$ with elements

$$T_{a,a'} = \operatorname{Tr}\left[M^{(a)} M^{(a')}\right]. \tag{7.9}$$

The POVM can be chosen such that $T$ is invertible and the POVM is called symmetric if the elements of $T$ are constant between distinct off-diagonal elements. As the POVM elements are factorable, $T$ is factorable as well and with this also $T^{-1}$ [1].

Assuming that $T$ is invertible, we can calculate the coefficients $Q_\rho(a')$ as

$$Q_\rho\left(a'\right) = \sum_{\{a\}} P\left(a\right) T_{a,a'}^{-1}, \tag{7.10}$$

so that we get an expression for the density matrix in terms of POVM elements [1],

$$\rho = \sum_{\{a\}} \sum_{\{a'\}} P\left(a\right) M^{(a')} T_{a,a'}^{-1}. \tag{7.11}$$

This enables the calculation of expectation values of general operators $O$ defined in the spin basis using the representation in terms of POVM elements,

$$\begin{aligned}
\langle O \rangle &= \text{Tr}\left[O\rho\right] \\
&= \sum_{\{a\}} \sum_{\{a'\}} P\left(a\right) T_{a,a'}^{-1} \text{Tr}\left[OM^{(a')}\right].
\end{aligned} \tag{7.12}$$

The trace can be rewritten as

$$\text{Tr}\left[OM^{(a')}\right] = \sum_{\{a''\}} Q_O\left(a''\right) T_{a',a''}, \tag{7.13}$$

by expanding $O$ in the POVM representation according to Eq. (7.6). The expectation value is thus given by [1]

$$\langle O \rangle = \sum_{\{a\}} P\left(a\right) Q_O\left(a\right), \tag{7.14}$$

using

$$\sum_{\{a'\}} T_{a,a'}^{-1} T_{a',a''} = \delta_{a,a''}. \tag{7.15}$$

The coefficients $Q_O(a)$ need to be calculated for each operator $O$ individually. Since we assume $T$ to be invertible, this can be done via

$$Q_O\left(a\right) = \sum_{\{a'\}} \text{Tr}\left[OM^{(a')}\right] T_{a',a}^{-1}. \tag{7.16}$$

The POVM elements are factorable, so that the coefficients can be calculated for single spin operators when considering locally acting operators, such as Pauli strings, which are used most commonly. The total coefficients are then the products of the individual single-site coefficients. Thus, expectation values of local operators can be evaluated efficiently and we do not have to back-transform the state from the POVM representation to the spin basis and create the full density matrix [1].

### 7.1.2 The Tetrahedral POVM

One possible choice to represent a spin-1/2 particle with a POVM which is informationally complete, symmetric and has an invertible overlap matrix is the tetrahedral POVM $\mathbf{M}_{\text{tetra}}$. The corresponding POVM elements project a single spin onto the corners of a tetrahedron lying in the Bloch sphere, as depicted in Fig. 7.1. The choice of the tetrahedron is not unique, as it can be rotated inside the Bloch sphere. Also, the tetrahedral POVM is only one possible choice of a POVM with the properties stated above, more possible choices can exist [1, 2].

In the following, we focus on the tetrahedral POVM, which is given for a single spin by [1, 2]

$$\mathbf{M}_{\text{tetra}} = \left\{ M^{(a)} = \frac{1}{4}\left(\mathbb{1} + \mathbf{s}^{(a)}\boldsymbol{\sigma}\right)\right\}_{a \in \{0,1,2,3\}}, \tag{7.17}$$

$$\mathbf{s}^{(0)} = (0, 0, 1),$$

$$\mathbf{s}^{(1)} = \left(\frac{2\sqrt{2}}{3}, 0, -\frac{1}{3}\right),$$

$$\mathbf{s}^{(2)} = \left(-\frac{\sqrt{2}}{3}, \sqrt{\frac{2}{3}}, -\frac{1}{3}\right), \tag{7.18}$$

$$\mathbf{s}^{(3)} = \left(-\frac{\sqrt{2}}{3}, -\sqrt{\frac{2}{3}}, -\frac{1}{3}\right),$$

with Pauli matrices $\boldsymbol{\sigma} = (\sigma^x, \sigma^y, \sigma^z)$. The POVM elements take the form

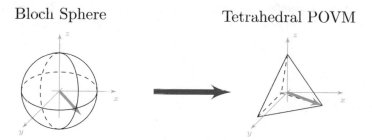

Bloch Sphere          Tetrahedral POVM

**Fig. 7.1** Illustrative projection of the state space of a single spin-1/2 particle from the Bloch sphere onto the tetrahedral POVM [Eq. (7.17)]. While in the spin basis the state can be described by any point on the surface of the Bloch sphere, where it can take the two values $\pm 1$ in the $z$-direction and has an additional complex phase according to the coordinates in $x$- and $y$-direction, it can take one of the four possible states at the corners of the tetrahedron spanned by the measurement outcomes of the POVM elements. Each corner corresponds to one measurement outcome. The representation of one possible spin configuration is illustrated by the arrow

$$M^{(0)} = \frac{1}{2}\begin{pmatrix} 1 & 0 \\ 0 & 0 \end{pmatrix},$$

$$M^{(1)} = \frac{1}{6}\begin{pmatrix} 1 & \sqrt{2} \\ \sqrt{2} & 2 \end{pmatrix},$$

$$M^{(2)} = \frac{1}{12}\begin{pmatrix} 2 & -\sqrt{2} - \sqrt{6}i \\ -\sqrt{2} + \sqrt{6}i & 4 \end{pmatrix},$$

$$M^{(3)} = \frac{1}{12}\begin{pmatrix} 2 & -\sqrt{2} + \sqrt{6}i \\ -\sqrt{2} - \sqrt{6}i & 4 \end{pmatrix},$$

(7.19)

giving the overlap matrix

$$T = \frac{1}{4}\begin{bmatrix} 1 & 1/3 & 1/3 & 1/3 \\ 1/3 & 1 & 1/3 & 1/3 \\ 1/3 & 1/3 & 1 & 1/3 \\ 1/3 & 1/3 & 1/3 & 1 \end{bmatrix},$$

(7.20)

with inverse

$$T^{-1} = \begin{bmatrix} 5 & -1 & -1 & -1 \\ -1 & 5 & -1 & -1 \\ -1 & -1 & 5 & -1 \\ -1 & -1 & -1 & 5 \end{bmatrix}.$$

(7.21)

The POVM elements for systems of multiple spin-1/2 particles can be calculated via the tensor product of the single-particle POVM elements, which is analogously true for the overlap matrix and its inverse [1].

To perform measurements, we can use Eq. (7.14), where we calculate the coefficients $Q_O(a)$ for a specific operator $O$ acting on a single spin according to Eq. (7.16) using the tetrahedral POVM [1]. Considering the $\sigma^z$-operator,

$$\sigma^z = \begin{bmatrix} 1 & 0 \\ 0 & -1 \end{bmatrix},$$

(7.22)

we get the coefficients

$$Q_{\sigma^z}(0) = 3,$$
$$Q_{\sigma^z}(1) = -1,$$
$$Q_{\sigma^z}(2) = -1,$$
$$Q_{\sigma^z}(3) = -1.$$

(7.23)

This gives

$$\langle \sigma^z \rangle = 3P\,(a=0) - P\,(a=1) - P\,(a=2) - P\,(a=3).$$

(7.24)

Analogously, we can consider the $\sigma^x$-operator,

$$\sigma^x = \begin{bmatrix} 0 & 1 \\ 1 & 0 \end{bmatrix}, \tag{7.25}$$

which provides the coefficients

$$
\begin{aligned}
Q_{\sigma^x}(0) &= 0, \\
Q_{\sigma^x}(1) &= 2\sqrt{2}, \\
Q_{\sigma^x}(2) &= -\sqrt{2}, \\
Q_{\sigma^x}(3) &= -\sqrt{2},
\end{aligned}
\tag{7.26}
$$

so that we obtain

$$\langle \sigma^x \rangle = 2\sqrt{2}P\,(a = 1) - \sqrt{2}P\,(a = 2) - \sqrt{2}P\,(a = 3). \tag{7.27}$$

Given these two expressions, any Pauli string operator which is a combination of the two operators acting on multiple spin particles can be measured, since the $Q_O(a)$ factorize [1].

Taking a closer look at the calculated coefficients, these can reach values which are not possible when considering measurements in the spin basis. As an example we assume the single-particle state represented by a tetrahedral POVM with

$$
\begin{aligned}
P\,(a = 0) &= 1, \\
P\,(a = 1) = P\,(a = 2) = P\,(a = 3) &= 0.
\end{aligned}
\tag{7.28}
$$

Performing a measurement with $\sigma^z$ yields the expectation value

$$\langle \sigma^z \rangle = 3. \tag{7.29}$$

In the spin basis, this measurement can only give two possible outcomes,

$$
\begin{aligned}
\sigma^z |\uparrow\rangle &= 1, \\
\sigma^z |\downarrow\rangle &= -1,
\end{aligned}
\tag{7.30}
$$

so that

$$\left| \langle \sigma^z \rangle \right| \leq 1. \tag{7.31}$$

Thus, the state represented by the POVM is not a physical state. We can calculate its density matrix using Eq. (7.11), which gives

$$\rho = \begin{bmatrix} 2 & 0 \\ 0 & -1 \end{bmatrix}. \tag{7.32}$$

This matrix is not positive definite, which is however a property of the density matrix. Thus, we can conclude that all physical states can be represented by the tetrahedral POVM since it is informationally complete. Nevertheless, not all probability distributions underlying the measurement outcomes of the tetrahedral POVM elements represent physical states. The state space in the POVM representation is hence larger than the Hilbert space in the spin basis.

Further constraints on the probabilities $P(\boldsymbol{a})$ would be necessary to restrict the state space to physical states, but these constraints require knowledge of the full density matrix. Thus, they are exponentially hard to calculate for large spin systems, so that they cannot be applied efficiently. However, it can straightforwardly be shown that the density matrices corresponding to states represented by the tetrahedral POVM are always Hermitian and have trace one. Hence, the positive definiteness is the only property of the density matrix which is not generally fulfilled.

### 7.1.3 Representing Spin States with a Real-Valued RBM Using POVM

A spin state expressed in the POVM representation can be fully described using the real probabilities $P(\boldsymbol{a})$ defined by the POVM elements. The corresponding probability distribution can directly be represented by an RBM with real-valued weights and biases [1]. Thus, the spin state can be parametrized by a real-valued RBM. Analogously to the RBM parametrization introduced in Sect. 5.1 the visible variables of the real-valued RBM then correspond to the states in the POVM representation, which we index by $\boldsymbol{a}$. Each element $a_i$ can take four possible values, $a_i \in \{0, 1, 2, 3\}$, so that we have to modify the RBM ansatz to get non-binary visible variables. This is in general possible, since in a classical RBM with binary variables one can always combine two of those to a single neuron, which can then take four possible states. Therefore, twice as many visible neurons are required [4, 5].

While this is a general argument how to encode four-state variables in a binary RBM, in the software simulations this can be done more efficiently by using the so-called one-hot encoding scheme, where each visible variable corresponds to a vector $\mathbf{v}_i = (v_i^0, v_i^1, v_i^2, v_i^3)$ with four binary entries, $v_i^k \in \{0, 1\}$. Each entry then denotes whether the neuron is in one of the four possible states, which leads to the restriction that exactly one of the four entries is one, denoting the state the neuron is in, while the remaining three are zero [1].

We use this one-hot encoding for the simulations in the following. The relation between visible variables $\mathbf{v}_i$ and variables $a_i$ indexing the configuration in the POVM representation is then given by

$$v_i^k = \begin{cases} 1 \text{ if } k = a_i, \\ 0 \text{ else,} \end{cases} \tag{7.33}$$

with $k \in \{0, 1, 2, 3\}$. The other way around we get the relation

$$a_i = v_i^1 + 2v_i^2 + 3v_i^3. \tag{7.34}$$

Even though the hidden variables are still binary, we encode them via the one-hot scheme similar to the visible variables to keep the ansatz consistent. The one-hot encoded hidden neurons are then vectors with two elements where exactly one is zero and the other is one. The translation from the binary state into the one-hot encoding works similarly as for the visible variables.

The energy of the adapted RBM takes a slightly different form, since the connecting weights and biases can take different values depending on which state the visible variable is in. Thus, each connecting weight takes four indices, two indicating which visible and which hidden neuron it connects and two indicating in which state the visible and the hidden neuron are, respectively. The biases need to be adapted similarly, so that each bias takes two indices [1].

The energy of an RBM with $N$ visible and $M$ hidden variables in the one-hot encoding is then given by

$$E(v, h; \mathcal{W}) = -\sum_{i=1}^{N}\sum_{j=1}^{M}\sum_{k=0}^{3}\sum_{l=0}^{1} v_i^k W_{i,j}^{k,l} h_j^l - \sum_{i=1}^{N}\sum_{k=0}^{3} v_i^k d_i^k - \sum_{j=1}^{M}\sum_{l=0}^{1} h_j^k b_j^k. \tag{7.35}$$

The indices $k$ and $l$ run over the four or two elements of each visible or hidden neuron, respectively. There is always exactly one of these elements which is one and the others are zero, indicating which state the neuron is in [1]. Thus, the sets of all visible and hidden neurons are given by matrices $v = (\mathbf{v}_1, \ldots, \mathbf{v}_N)$, $h = (\mathbf{h}_1, \ldots, \mathbf{h}_M)$, respectively. The corresponding network setup is visualized in Fig. 7.2.

The unnormalized probability distribution underlying the visible variables in the RBM is then given by

$$P(v; \mathcal{W}) = \sum_{\{h\}} \exp\left[-E(v, h; \mathcal{W})\right], \tag{7.36}$$

and can be normalized with the factor

$$Z(\mathcal{W}) = \sum_{\{h\}}\sum_{\{v\}} \exp\left[-E(v, h; \mathcal{W})\right]. \tag{7.37}$$

The sums run over all possible states of visible and hidden neurons.

Since we now consider a real-valued RBM, standard procedures from machine learning such as the contrastive divergence scheme, $CD_1$, as introduced in Sect. 3.3.2

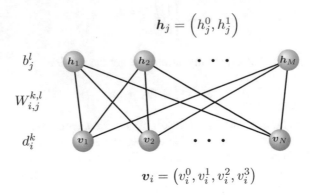

**Fig. 7.2** Illustration of the real-valued RBM representing the probability distribution underlying a spin state in the POVM representation. The visible variables can take four possible values which are implemented using a one-hot encoding, so that each variable is represented by a four-element vector with exactly one entry being one and the others zero. The hidden variables are binary but are also implemented via the one-hot encoding. Due to this encoding the weights and biases get extra indices stating which element of the variable vectors is considered

can be used to train the network [1]. Since the network is supposed to represent the probability distribution $P(a)$ underlying the states in the POVM representation, we need to translate this distribution into the one underlying the one-hot encoded visible variables of the RBM, $P(v)$. To do so, we use the relation

$$P\left(v_1^{l_1} = 1, v_2^{l_2} = 1, \ldots, v_N^{l_N} = 1\right) = P\left(a_1 = l_1, a_2 = l_2, \ldots, a_N = l_N\right), \quad (7.38)$$

with $l_i \in \{0, 1, 2, 3\}$. Here we only denote the variable vector entries which are one in the argument on the left-hand side, implying that the remaining three entries are zero.

Furthermore, to sample state configurations of visible and hidden variables in the RBM, standard Gibbs sampling as introduced in Sect. 3.2.2 can be applied. Here we also have to keep in mind that we use the one-hot encoding. This slightly changes the conditional probabilities from which the visible and hidden configurations are drawn via importance sampling. The conditional probability for a visible variable $v_i$, $P(v_i^k = 1|h; \mathcal{W})$, is defined for all four possible states, $k \in \{0, 1, 2, 3\}$, via

$$P\left(v_i^k = 1 \mid h; \mathcal{W}\right) = \frac{\exp\left[\sum_{j=1}^{M} \sum_{l=0}^{1} W_{i,j}^{k,l} h_j^l + d_i^k\right]}{\sum_{m=0}^{3} \exp\left[\sum_{j=1}^{M} \sum_{l=0}^{1} W_{i,j}^{m,l} h_j^l + d_i^m\right]}. \quad (7.39)$$

From the corresponding probability distribution, a state can be sampled for each visible variable, providing a full configuration $v$. The hidden variables are still binary and the underlying conditional probabilities are given by

$$P\left(h_j^l = 1 \mid v; \mathcal{W}\right) = \frac{\exp\left[\sum_{i=1}^{N} \sum_{k=0}^{3} v_i^k W_{i,j}^{k,l} + b_j^l\right]}{\sum_{m=0}^{1} \exp\left[\sum_{i=1}^{N} \sum_{k=0}^{3} v_i^k W_{i,j}^{k,m} + b_j^m\right]}. \qquad (7.40)$$

Given these expressions, the Gibbs sampling can be performed according to the binary scheme in Sect. 3.2.2.

We still need to keep in mind that the POVM representation can also yield unphysical states, as discussed earlier. Therefore, we cannot use this ansatz to find ground states by training the real-valued RBM via energy minimization as introduced in Sect. 5.1.2. This scheme can end up in states which have minimum energy but are not valid states in the spin system, as they do not have a positive definite density matrix. However, we can use this real-valued RBM parametrization to learn the representation of a given density matrix in a representation with only real coefficients.

The ansatz can hence be used for quantum state tomography, where the state of a system is represented in the RBM parametrization trained on some set of measurement data. Further samples according to the underlying distribution can then be created. This is for example helpful when training the network on experimental data, since then experimentally unachievable operators can be measured. As a real-valued RBM is used here, the sampling can be implemented on the neuromorphic hardware, which can sample efficiently from Boltzmann distributions. Since such a distribution underlies the visible and hidden variables in the RBM parametrization of spin states in the POVM representation, an implementation is straightforward and we expect a speedup in the sampling process from it [6–8].

## 7.2  Simulating Entangled Bell States Using Classical Networks

### 7.2.1  Tetrahedral POVM Representation of the Bell State

As a first example for simulating the RBM parametrization of a spin system in the POVM basis we consider the Bell state,

$$|\Psi^{\mathrm{BP}}\rangle = \frac{1}{\sqrt{2}} \left(|\uparrow\downarrow\rangle + |\downarrow\uparrow\rangle\right). \qquad (7.41)$$

This is a strongly entangled state and we can thus check whether the CHSH-inequality (named after Clauser, Horne, Shimony and Holt) can be violated in simulations based on a classical network. The Bell state is a two-spin system, so it requires $4^2 = 16$ POVM elements. We use a tetrahedral POVM for each spin and since there are not many elements we can calculate the probability distribution $P(\boldsymbol{a})$ analytically using Eq. (7.3).

The density matrix of the Bell state can be calculated from the state vector,

$$\rho = |\Psi^{\mathrm{BP}}\rangle\langle\Psi^{\mathrm{BP}}|$$

$$= \begin{bmatrix} 0 & 0 & 0 & 0 \\ 0 & 1/2 & 1/2 & 0 \\ 0 & 1/2 & 1/2 & 0 \\ 0 & 0 & 0 & 0 \end{bmatrix}. \tag{7.42}$$

Using the tetrahedral POVM elements, Eq. (7.19), and the property that they factorize,

$$M^{(a_1,a_2)} = M^{(a_1)} \otimes M^{(a_2)}, \tag{7.43}$$

we can calculate the probabilities $P(\boldsymbol{a})$ of the 16 POVM basis states indexed by $\boldsymbol{a} = (a_1, a_2), a_1, a_2 \in \{0, 1, 2, 3\}$, via Eq. (7.3). For convenience sake we express the result in a matrix $\mathbb{P}$ with elements $\mathbb{P}_{a_1,a_2} = P(\boldsymbol{a} = [a_1, a_2])$ and get

$$\mathbb{P} = \begin{bmatrix} 0 & 1/12 & 1/12 & 1/12 \\ 1/12 & 1/9 & 1/36 & 1/36 \\ 1/12 & 1/36 & 1/9 & 1/36 \\ 1/12 & 1/36 & 1/36 & 1/9 \end{bmatrix}. \tag{7.44}$$

We can now create a real-valued RBM according to Fig. 7.2 and train it to represent this probability distribution after translating it into the language of one-hot encoded variables using Eq. (7.38).

## 7.2.2   Training the RBM

To train the real-valued RBM we can use the contrastive divergence scheme, $\mathrm{CD}_1$, as introduced in Sect. 3.3.2, since the network learns to represent the probability distribution underlying a given input data set. To create the input data, we apply a Metropolis-Hastings sampling scheme as introduced in Sect. 5.1.5 to sample from the exact distribution $P(\boldsymbol{a})$. Every component $a_i$ is now not binary but can take one of four possible values. Thus, the proposed state $\tilde{\boldsymbol{a}}$ in the Metropolis-Hastings sampling is created by choosing a random component $a_i$ in the given state $\boldsymbol{a}$ and changing it to one of the remaining three possible values. This value is chosen randomly, where all three possibilities are equally probable. The new configuration is accepted with probability $A(\boldsymbol{a}, \tilde{\boldsymbol{a}}) = \min[1, P(\tilde{\boldsymbol{a}})/P(\boldsymbol{a})]$ or otherwise rejected. The sampled configuration is then added to the sample set.

Given this input data set we can apply $\mathrm{CD}_1$ to update the weights in the real-valued RBM. This training scheme requires samples of the visible and hidden variables drawn from the conditional probabilities, according to Gibbs sampling introduced in Sect. 3.2.2. Here we use the conditional probabilities as stated in Eqs. (7.39)–(7.40) for the one-hot encoded variables.

The training procedure then takes the following form,

1. use Metropolis-Hastings sampling to create a data set $S$ of measurement outcomes of the POVM elements according to the target distribution $P(a)$,
2. apply $CD_1$ to update the weights,

    a. choose a mini-batch of configurations $a$ from $S$,
    b. transform each configuration $a$ in the mini-batch into the one-hot encoded form, yielding the visible variables $v$, and perform one Gibbs sampling step to create a sample of hidden variables $h$,
    c. for each configuration $h$ perform one Gibbs sampling step to create a sample of visible variables $\tilde{v}$ given the configuration $h$,
    d. for each configuration $\tilde{v}$ perform one Gibbs sampling step to create a sample of hidden variables $\tilde{h}$ given the configuration $\tilde{v}$,
    e. calculate the weight update, $\Delta W_{i,j}^{k,l} = -\langle v_i^k h_j^l \rangle + \langle \tilde{v}_i^k \tilde{h}_j^l \rangle$, with analogous expressions for the biases, where the average is taken over all configurations resulting from the elements in the mini-batch,
    f. update the weights according to $W_{i,j}^{k,l} \rightarrow W_{i,j}^{k,l} - \varepsilon \Delta W_{i,j}^{k,l}$, and analogously for the biases,
    g. repeat from a) until all mini-batches have been taken into account,

3. repeat from 1. until the weights converge.

Once the network is trained, or even during the training procedure, expectation values of operators can be calculated. For this, a sample set of $R$ visible configurations $v_r$ is created using multiple Gibbs sampling steps and the considered operator is evaluated at each sample. Averaging over the whole sample set then approximates the sum over all configurations in Eq. (7.14),

$$\langle O \rangle \approx \frac{1}{R} \sum_{r=1}^{R} \sum_{k=0}^{3} \sum_{l=0}^{3} Q_O \left( v_{1,r}^k, v_{2,r}^l \right) v_{1,r}^k v_{2,r}^l. \tag{7.45}$$

We have substituted the variables $a$ in Eq. (7.14) by the one-hot encoded variables $v$. The lower index $r$ denotes the samples.

We have implemented the training procedure of a real-valued RBM to represent a Bell state and the results are shown in Fig. 7.3. We plot there the magnetizations of the first spin and the correlations between the two spins in the $x$- and $z$-basis as functions of the training epochs. One epoch refers to one full $CD_1$ step, corresponding to point 2. in the training procedure stated above. Furthermore, we plot the CHSH-observable

$$\mathcal{B}^{\mathrm{BP}} = \sqrt{2} \left[ \langle \sigma_1^x \sigma_2^x \rangle - \langle \sigma_1^z \sigma_2^z \rangle \right], \tag{7.46}$$

and the Kullback-Leibler divergence

$$\Xi(\mathcal{W}) = \sum_{\{v\}} P(v) \log\left[\frac{P(v)\,Z(\mathcal{W})}{P(v;\mathcal{W})}\right], \tag{7.47}$$

with normalization factor $Z(\mathcal{W})$ as stated in Eq. (7.37) and unnormalized distribution $P(v;\mathcal{W})$ as stated in Eq. (7.36). These quantities are also plotted as functions of the training epochs and all results are compared with the expected solutions. We vary the number of hidden variables in the network from $M = 1$ to $M = 5$ in steps of one to find convergence towards the exact results with increasing $M$.

The simulations are done on a training data set containing $10^5$ samples drawn via Metropolis-Hastings sampling from the probability distribution representing the Bell state, Eq. (7.44). The weight updates are calculated on this training data during each epoch using the $CD_1$ scheme. We divide the whole training data into mini-batches containing 10 samples each. To calculate the expectation values, $10^5$ samples are drawn from the Boltzmann distribution underlying the RBM via block Gibbs sampling. The network is trained ten times for each value of $M$ and the results are

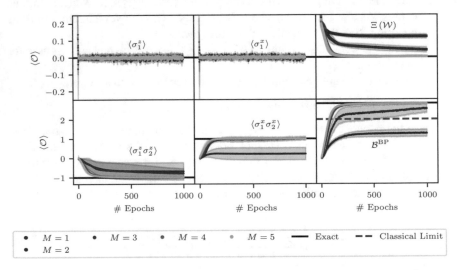

**Fig. 7.3** Training process of the RBM learning to represent a Bell state in the POVM representation. Magnetizations of the first spin ($\langle\sigma_1^x\rangle$, $\langle\sigma_1^z\rangle$), as well as correlations between the two spins ($\langle\sigma_1^x\sigma_2^x\rangle$, $\langle\sigma_1^z\sigma_2^z\rangle$) in the $x$- and $z$-basis are plotted as functions of training epochs in the $CD_1$ scheme. Furthermore, the convergence of the Kullback-Leibler divergence $\Xi(\mathcal{W})$ [Eq. (7.47)] and of the CHSH-observable $\mathcal{B}^{BP}$ [Eq. (7.46)] is shown. The number of hidden variables is varied from $M = 1$ to $M = 5$ in steps of one, indicated by the different colors. For each $M$ the network is trained ten times from random initial weights. The results are averaged, and shaded regions denote statistical fluctuations. For each observable the expected value is indicated by the black line and in the lower right plot the red dashed line denotes the classical limit of the CHSH-inequality, indicating its violation. The exact solution is found to be represented well for $M \geq 3$ and the Bell inequality is violated. A Kullback-Leibler divergence around $\Xi(\mathcal{W}) \lesssim 0.002$ appears to indicate a representation of the Bell state with high accuracy

averaged with the shaded regions denoting statistical fluctuations. The learning rate is set to $\varepsilon = 10^{-4}$.

We find good convergence for all observables for $M \geq 3$ in Fig. 7.3, while the expected correlations are not reached for $M < 3$ and even the Kullback-Leibler divergence remains larger for these cases. This shows that a Kullback-Leibler divergence of $\Xi(W) \lesssim 0.002$ is necessary to capture the expected correlations. The magnetizations are always represented with good accuracy. The CHSH-inequality is violated for $M \geq 2$ already, even though the solution does not fully reach the exact values for $M = 2$. However, full convergence is not yet reached for $M = 2$ and the exact solution might be captured more accurately after more training epochs. Nevertheless, calculating more training epochs requires more computation time than increasing the number of hidden variables here, so that we do not analyze this behavior further.

With this we demonstrate that the POVM ansatz can indeed be used to represent a Bell state given a sufficient amount of hidden neurons, which is in accordance with results in [1]. From these observations we expect that this strongly entangled state can also be simulated on the neuromorphic hardware providing a speedup in the sampling procedure. However, the weights still need to be trained on a classical computer, as discussed in Sect. 3.5.

### 7.2.3 Outlook: Building the Way Towards Neuromorphic Hardware

As we have found an approach to represent a Bell state with a purely real-valued RBM, the next step is to bring the sampling procedure onto the neuromorphic hardware of the BrainScaleS group as introduced in Sect. 3.5. We can then compare software and hardware simulations. On the way to simulations on the neuromorphic chips, several differences to the software implementation as used in Sect. 7.2.2 arise.

First, the leaky integrate-and-fire (LIF) neurons can only represent binary variables, since the sampling is defined via the ratio of a spiking neuron being in a refractory state or not, as discussed in Sect. 3.5 [6–8]. Thus, on the hardware we need twice as many visible variables compared to the software implementation and each two visible neurons are combined to represent one four-state neuron. While this translation works in principle it yields a different representation. Thus, the two implementations converge to different weights and biases and it is not clear if the same number of hidden neurons is sufficient for both cases and how well the convergence performs.

Furthermore, the weights on the neuromorphic hardware in the BrainScaleS group can only take discretized values as they can only be implemented in a four-bit representation. This limits the representational power of the RBM, but we still expect that the desired probability distribution can be represented with sufficient accuracy, given that we choose enough hidden variables [9]. As only the sampling is performed with the spiking neurons, the training of the network can still be done via contrastive divergence, $CD_1$, just as in the software simulation. However, the learning rate can differ between the two implementations and thus needs to be adapted.

According to previous applications of the BrainScaleS hardware [6–8] we do not expect these differences to cause any problems, as the network can be adapted accordingly. We still expect a speedup in the sampling [10]. However, we have found that a small Kullback-Leibler divergence is necessary, $\Xi(\mathcal{W}) \lesssim 0.002$, to capture the correlations accordingly. It is not clear whether such a high precision in representing a probability distribution is reachable on the neuromorphic hardware, as for other applications it has not been necessary yet [8].

The implementation itself is still outstanding, but we have provided a way to sample entangled two-particle quantum states on classical analog networks, which can straightforwardly and efficiently be generalized to larger system sizes. With this we provide an ansatz to perform quantum state tomography on neuromorphic hardware.

# References

1. Juan C, Giacomo T, Melko Roger G, Leandro A (2019) Reconstructing quantum states with generative models. Nat Mach Intell 1(3):155–161. https://doi.org/10.1038/s42256-019-0028-1

2. Tabia GNM (2012) Experimental scheme for qubit and qutrit symmetric informationally complete positive operator-valued measurements using multiport devices. Phys Rev A 86:062107. https://link.aps.org/doi/10.1103/PhysRevA.86.062107

3. Nielsen MA, Chuang IL (2010) Quantum computation and quantum information: 10th, anniversary edition. Cambridge University Press. https://doi.org/10.1017/CBO9780511976667

4. Hinton GE (2012) A practical guide to training restricted Boltzmann machines. Springer, Berlin, pp 599–619. https://doi.org/10.1007/978-3-642-35289-8_32

5. Guido M, Jason M (2015) Discrete restricted Boltzmann machines. J Mach Learn Res 16:653–672. http://jmlr.org/papers/v16/montufar15a.html

6. Petrovici MA (2016) Form versus function: theory and models for neuronal substrates. Springer International Publishing, Berlin. https://doi.org/10.1007/978-3-319-39552-4

7. Petrovici MA, Johannes B, Ilja B, Johannes S, Karlheinz M (2016) Stochastic inference with spiking neurons in the high-conductance state. Phys Rev E 94:042312. https://link.aps.org/doi/10.1103/PhysRevE.94.042312

8. Kungl AF, Schmitt S, Klähn J, Müller P, Baumbach A, Dold D, Kugele A, Müller E, Koke C, Kleider M, Mauch C, Breitwieser O, Leng L, Gürtler N, Güttler M, Husmann D, Husmann K, Hartel A, Karasenko V, Grübl A, Schemmel J, Meier K, Petrovici Mihai A (2019) Accelerated physical emulation of bayesian inference in spiking neural networks. Front. Neurosci 13:1201. https://www.frontiersin.org/article/10.3389/fnins.2019.01201

9. Rastegari M, Ordonez V, Redmon J, Farhadi A (2016) XNOR-net: ImageNet classification using binary convolutional neural networks. arXiv:1603.05279 [cs.CV]

10. Wunderlich T, Kungl AF, Müller E, Hartel A, Stradmann Y, Aamir SA, Grübl A, Heimbrecht A, Schreiber K, Stöckel D, Pehle C, Billaudelle S, Kiene G, Mauch C, Schemmel J, Meier K, Petrovici MA (2019) Demonstrating advantages of neuromorphic computation: a pilot study. Front Neurosci 13:260. https://www.frontiersin.org/article/10.3389/fnins.2019.00260

# Chapter 8
# Conclusion

In this thesis we have studied approximate simulation methods of ground states and dynamics in quantum many-body systems, specifically spin-1/2 systems, where general efficient simulation techniques are still missing. We have considered the discrete truncated Wigner approximation (dTWA) as a semi-classical phase-space method [1–3], as well as an approach based on quantum Monte Carlo techniques. The latter uses a parametrization of quantum state vectors in terms of a complex-valued restricted Boltzmann machine (RBM) [4]. We have benchmarked the two simulation methods on dynamics after sudden quenches in the transverse-field Ising model (TFIM) by comparison with analytical results. The TFIM defines a quantum phase transition between a paramagnetic and a ferromagnetic phase. We have simulated sudden quenches from an effectively fully polarized initial state deep in the paramagnetic phase to different distances from the quantum critical point. Here we have considered both, quenches within one regime and quenches across the quantum phase transition [5–7].

The equilibrated correlation function after a sudden quench within the paramagnetic phase in the TIFM shows an exponential decay as a function of the relative distance between the considered spins. From this decay, two correlation lengths can be extracted. The first correlation length is found at short relative distances for all quenches, while the second appears at larger relative distances and only for quenches to larger distances from the quantum phase transition. We have studied whether the two simulation techniques can capture these correlation lengths in the long-time limit [5–7].

In Chap. 4 we have analyzed the performance of the dTWA with truncation at first and second order. We found that at short times the dynamics are always captured well, and the second-order simulation performs more accurately. However, at later times the second-order equations of motion experience instabilities leading to divergences in the simulations. The truncation at first order remains stable, so that

S. Czischek, *Neural-Network Simulation of Strongly Correlated Quantum Systems*, Springer Theses, https://doi.org/10.1007/978-3-030-52715-0_8

we have studied it in more detail. We have, however, found deviations from the exact solution appearing at later times. These deviations we found especially for quenches to intermediate distances from the quantum critical point. The short-distance correlation length we found to be captured with good accuracy close to and far away from the quantum critical point. At intermediate distances from the quantum phase transition the correlation length saturates at different long-time limits. The second correlation length was not captured at all in the dTWA simulations.

In Chap. 5 we have analogously benchmarked the RBM parametrization approach on sudden quenches in the TFIM and have given a direct comparison with the dTWA results. We have found that the exact solution can be captured accurately with this ansatz if only enough free parameters, which are adapted variationally, are chosen in the parametrization. However, when analyzing the scaling of the necessary number of variational parameters with system size we have experienced that the Hilbert space dimension needs to be fully covered for quenches into the quantum critical regime. Further away from the quantum phase transition a small number of variational parameters turned out to be enough. Thus, we found this ansatz to be inefficient for quenches close to the quantum critical point, where strong long-range correlations and volume-law entanglement are expected. In those regimes the number of variational parameters scales exponentially with the system size.

We have furthermore added a longitudinal field to the TFIM and have benchmarked the two simulation methods in this non-integrable model [8]. Here we have again considered sudden quenches from an effectively fully polarized state to different final fields and have experienced the same behavior as for quenches in the TFIM. We have compared the simulation results with exact diagonalization for small system sizes or with time-dependent density-matrix renormalization group (tDMRG) for lager system sizes. The latter is known to describe the exact solution well [9, 10]. We have found that the dTWA captures the short-time dynamics accurately but shows deviations in all regimes at later times. The RBM parametrization ansatz has shown deviations in regimes of large correlation lengths, which could be minimized by increasing the number of variational parameters. However, this leads to an exponential scaling of the number of variational parameters with the system size, rendering the ansatz inefficient in regimes of strong long-range correlations. These are regimes where also other existing simulation methods, such as tDMRG, struggle and behave inefficiently. We have furthermore demonstrated that the performance of the RBM ansatz is comparable with tDMRG results when choosing an approximately equal number of variational parameters. Since the tDMRG method can be implemented much more efficiently, the RBM parametrization stands behind it, as it struggles in the same regimes. Nevertheless, this method can be applied in higher dimensional systems where a similar performance is expected. This is not possible for tDMRG methods.

Summing up, we have pointed out limitations of approximate simulation methods of quantum spin-1/2 systems on classical computers in regimes of strong long-range correlations. This motivated us to study whether computing devices going beyond the von-Neumann architecture can help to circumvent these limitations. For this we have considered the neuromorphic hardware in the BrainScaleS group at Heidelberg

University, which emulates a neural network with analog circuits and can efficiently create samples in an RBM [11–14]. This brought up the idea of combining the complex-valued RBM parametrization of quantum state vectors with the hardware, from which we expect an efficient way to create samples of spin states to approximately evaluate expectation values of quantum operators. Moreover, we hope to get further insights into quantum effects by the simulation of quantum states on a classical analog hardware. However, since the RBM used in the state vector parametrization is complex-valued, it cannot directly be implemented on the neuromorphic hardware but needs to be adapted first. We have analyzed two approaches which in principle enable such an implementation and have benchmarked their applicability on simulating entangled quantum states. We have analyzed their benefits and limitations and thus showed a way towards using the neuromorphic hardware for simulations of quantum many-body systems.

In Chap. 6 we have introduced the phase reweighting ansatz to sample spin states from the complex-valued RBM parametrization in a way that can also be implemented on the neuromorphic hardware. This ansatz absorbs the complex phases into the evaluated observables, so that samples are drawn from the Boltzmann distribution defined by the real parts of the weights in the RBM [15–18]. Additionally, this sampling scheme allows an extension to deep neural networks by adding hidden layers. This we have used to introduce a parametrization allowing an efficient evaluation of observables in any cartesian basis by only considering visible variables of the network. We have benchmarked this approach on the ground state in the TFIM at the quantum critical point, as well as on the strongly entangled Bell and Greenberger-Horne-Zeilinger (GHZ) states. We have found that for several states, especially for the strongly entangled ones, an exponential amount of samples is necessary to approximate expectation values of quantum operators with good accuracy. This phenomenon is known as the sign problem and it appears in phase reweighting approaches in quantum Monte Carlo schemes due to heavily fluctuating phase factors [15–20].

Considering operators which are diagonal in the spin basis parametrized via the shallow RBM, we have found that some states, like the ground state in the TFIM at the quantum critical point, can be sampled efficiently with the phase reweighting scheme. For these states the number of necessary samples to reach a certain accuracy is small and does not significantly depend on the system size. In contrast to this, when performing measurements in different cartesian bases parametrized via deep neural networks, we have always experienced the sign problem. Thus, the deep networks cannot be sampled efficiently as the necessary sample size to reach a certain accuracy scales exponentially with the system size. Even though this problem limits the simulation ansatz to small systems, we argue that the neuromorphic hardware can enable simulations of larger systems due to the efficient implementation of the sampling procedure. This enables a sampling of more spin states in shorter time. However, also with the help of the neuromorphic hardware the ansatz is still limited in the system size. The limit can at best be shifted to longer spin chains.

In Chap. 7 we have used a representation of quantum states via positive-operator valued measures (POVM), in which any quantum state can be expressed with purely non-negative basis-expansion coefficients [21]. Thus, this representation provides a real-valued probability distribution fully describing a spin state which can be represented by a real-valued RBM. The sampling from this network can directly be performed on the neuromorphic hardware. We have implemented an RBM in a standard software simulation and have trained it to represent a strongly entangled Bell state. We have analyzed the performance of this ansatz to get further insights into the requirements on the neuromorphic hardware to simulate such a state. Based on these observations we have discussed the upcoming steps towards an implementation on the BrainScaleS hardware as an outlook.

In summary we have experienced that quantum many-body systems cannot generally be simulated on classical computers in an efficient way, but most approximate simulation methods struggle in regimes of strong long-range correlations and quantum entanglement. We hence have started to devise ways towards implementing such simulations on neuromorphic hardware, analyzing whether this different architecture can help to overcome these problems. Here we have experienced further difficulties. As we have by far not found a general way to realize such an implementation, but only made first steps in this direction, a lot of future work remains open. While the POVM ansatz shows promising results, its performance still needs to be analyzed on the neuromorphic hardware and for more general systems. Moreover, a connection to experimental data, corresponding to quantum state tomography, yields an interesting task. An ansatz to find ground states or even simulate dynamics in this representation is as well still pending.

# References

1. Schachenmayer J, Pikovski A, Rey AM (2015) Dynamics of correlations in two-dimensional quantum spin models with long-range interactions: a phase-space Monte-Carlo study. New J Phys 17(6):065009. https://doi.org/10.1088/1367-2630/17/6/065009
2. Schachenmayer J, Pikovski A, Rey AM (2015) Many-body quantum spin dynamics with Monte Carlo trajectories on a discrete phase space. Phys Rev X 5:011022. https://link.aps.org/doi/10.1103/PhysRevX.5.011022
3. Pucci L, Roy A, Kastner M (2016) Simulation of quantum spin dynamics by phase space sampling of Bogoliubov-Born-Green-Kirkwood-Yvon trajectories. Phys Rev B 93(17):174302. https://link.aps.org/doi/10.1103/PhysRevB.93.174302
4. Carleo G, Troyer M (2017) Solving the quantum many-body problem with artificial neural networks. Science 355(6325):602–606. http://science.sciencemag.org/content/355/6325/602
5. Calabrese P, Essler FHL, Fagotti M (2012) Quantum quench in the transverse field Ising chain: I. time evolution of order parameter correlators. J Stat Mech: Theory Exp 2012(07):P07016. https://doi.org/10.1088%2F1742-5468%2F2012%2F07%2Fp07016
6. Calabrese P, Essler FHL, Fagotti M (2012) Quantum quenches in the transverse field Ising chain: II. stationary state properties. J Stat Mech: Theory Exp 2012(07):P07022. https://doi.org/10.1088%2F1742-5468%2F2012%2F07%2Fp07022
7. Karl M, Cakir H, Halimeh JC, Oberthaler MK, Kastner M, Gasenzer T (2017) Universal equilibrium scaling functions at short times after a quench. Phys Rev E 96:022110. https://link.aps.org/doi/10.1103/PhysRevE.96.022110

8. Ovchinnikov AA, Dmitriev DV, Krivnov VY, Cheranovskii VO (2003) Antiferromagnetic Ising chain in a mixed transverse and longitudinal magnetic field. Phys Rev B 68(21):214406. https://link.aps.org/doi/10.1103/PhysRevB.68.214406

9. Schollwöck U (2011) The density-matrix renormalization group in the age of matrix product states. Ann Phys 326(1):96–192. http://www.sciencedirect.com/science/article/pii/S0003491610001752

10. Bridgeman JC, Chubb CT (2017) Hand-waving and interpretive dance: an introductory course on tensor networks. J Phys A: Math Theor 50(22):223001. https://doi.org/10.1088%2F1751-8121%2Faa6dc3

11. Wunderlich T, Kungl AF, Müller E, Hartel A, Stradmann Y, Aamir SA, Grübl A, Heimbrecht A, Schreiber K, Stöckel D, Pehle C, Billaudelle S, Kiene G, Mauch C, Schemmel J, Meier K, Petrovici MA (2019) Demonstrating advantages of neuromorphic computation: a pilot study. Front Neurosci 13:260. https://www.frontiersin.org/article/10.3389/fnins.2019.00260

12. Petrovici MA (2016) Form versus function: theory and models for neuronal substrates. Springer International Publishing, Berlin. https://doi.org/10.1007/978-3-319-39552-4

13. Petrovici MA, Bill J, Bytschok I, Schemmel J, Meier K (2016) Stochastic inference with spiking neurons in the high-conductance state. Phys Rev E 94:042312. https://link.aps.org/doi/10.1103/PhysRevE.94.042312

14. Kungl AF, Schmitt S, Klähn J, Müller P, Baumbach A, Dold D, Kugele A, Müller E, Kokc C, Kleider M, Mauch C, Breitwieser O, Leng L, Gürtler N, Güttler M, Husmann D, Husmann K, Hartel A, Karasenko V, Grübl A, Schemmel J, Meier K, Petrovici MA (2019) Accelerated physical emulation of bayesian inference in spiking neural networks. Front Neurosci 13:1201. https://www.frontiersin.org/article/10.3389/fnins.2019.01201

15. Nakamura T, Hatano N, Nishimori H (1992) Reweighting method for quantum Monte Carlo simulations with the negative-sign problem. J Phys Soc Jpn 61(10):3494–3502. https://doi.org/10.1143/JPSJ.61.3494

16. Troyer M, Wiese U-J (2005) Computational complexity and fundamental limitations to fermionic quantum Monte Carlo simulations. Phys Rev Lett 94:170201.https://link.aps.org/doi/10.1103/PhysRevLett.94.170201

17. Anagnostopoulos KN, Nishimura J (2002) New approach to the complex-action problem and its application to a nonperturbative study of superstring theory. Phys Rev D 66:106008. https://link.aps.org/doi/10.1103/PhysRevD.66.106008

18. Loh EY, Gubernatis JE, Scalettar RT, White SR, Scalapino DJ, Sugar RL (1990) Sign problem in the numerical simulation of many-electron systems. Phys Rev B 41:9301–9307. https://link.aps.org/doi/10.1103/PhysRevB.41.9301

19. Torlai G, Carrasquilla J, Fishman MT, Melko RG, Fisher MPA (2019) Wavefunction positivization via automatic differentiation. arXiv:1906.04654 [quant-ph]

20. Hangleiter D, Roth I, Nagaj D, Eisert J (2019) Easing the Monte Carlo sign problem. arXiv:1906.02309 [quant-ph]

21. Carrasquilla J, Torlai G, Melko RG, Aolita L (2019) Reconstructing quantum states with generative models. Nat Mach Intell 1(3):155–161. https://doi.org/10.1038/s42256-019-0028-1

Printed in the United States
by Baker & Taylor Publisher Services